Emerging Technology Beyond 2035

Scenario-Based Technology Assessment for Future Military Contingencies

BRYAN BOLING, BENJAMIN BOUDREAUX, ALEXIS A. BLANC,
CHRISTY FORAN, EDWARD GEIST, MOON KIM, KELLY KLIMA,
ERIN N. LEIDY, SAMANTHA MCBIRNEY, DANIELLE C. TARRAF

Prepared for the United States Army
Approved for public release; distribution unlimited

RAND ARROYO CENTER

For more information on this publication, visit **www.rand.org/t/RRA1564-1**.

About RAND

The RAND Corporation is a research organization that develops solutions to public policy challenges to help make communities throughout the world safer and more secure, healthier and more prosperous. RAND is nonprofit, nonpartisan, and committed to the public interest. To learn more about RAND, visit www.rand.org.

Research Integrity

Our mission to help improve policy and decisionmaking through research and analysis is enabled through our core values of quality and objectivity and our unwavering commitment to the highest level of integrity and ethical behavior. To help ensure our research and analysis are rigorous, objective, and nonpartisan, we subject our research publications to a robust and exacting quality-assurance process; avoid both the appearance and reality of financial and other conflicts of interest through staff training, project screening, and a policy of mandatory disclosure; and pursue transparency in our research engagements through our commitment to the open publication of our research findings and recommendations, disclosure of the source of funding of published research, and policies to ensure intellectual independence. For more information, visit www.rand.org/about/research-integrity.

RAND's publications do not necessarily reflect the opinions of its research clients and sponsors.

Published by the RAND Corporation, Santa Monica, Calif.
© 2022 RAND Corporation
RAND® is a registered trademark.

Library of Congress Control Number: 2022914921
ISBN: 978-1-9774-0999-7

Cover Designer: Carol Ponce
Composite: Chris Desmond/U.S. Department of Defense; Gontar Alex/Getty Images.

About This Report

This report documents research and analysis conducted as part of a project entitled *Impact of Emerging Technology Trends Beyond 2035*, sponsored by Army Futures Command. The purpose of the project was to identify potential emerging technology trends beyond 2035, assess the feasibility of applying these emerging technologies to military operations, and describe the military implications of these technologies across the competition continuum.

This research was conducted within RAND Arroyo Center's Strategy, Doctrine, and Resources Program. RAND Arroyo Center, part of the RAND Corporation, is a federally funded research and development center (FFRDC) sponsored by the United States Army.

RAND operates under a "Federal-Wide Assurance" (FWA00003425) and complies with the *Code of Federal Regulations for the Protection of Human Subjects Under United States Law* (45 CFR 46), also known as "the Common Rule," as well as with the implementation guidance set forth in DoD Instruction 3216.02. As applicable, this compliance includes reviews and approvals by RAND's Institutional Review Board (the Human Subjects Protection Committee) and by the U.S. Army. The views of sources utilized in this study are solely their own and do not represent the official policy or position of DoD or the U.S. Government.

Acknowledgments

We would like to acknowledge the support of our project sponsors at the Army Futures Command's Futures and Concepts Center—specifically, MAJ Adam Taliaferro (our Action Officer) and LTC Keith Donnell. Their continued presence throughout project execution provided valuable feedback for the project team, and they facilitated data collection from across the Army Futures Command. We would also like to thank MAJ Matthew McDaniel for helping us understand the Army Futures Command science and technology portfolio and providing valuable discussion on future technology trends from the U.S. Army's perspective beyond 2035.

We would also like to thank our RAND colleagues who contributed to the success of this project. Jennifer Kavanagh and Stephan Watts provided valuable feedback from the RAND Arroyo Strategy, Doctrine, and Resources Program's perspective, and facilitated program support throughout project execution. Additionally, we would like to thank Laurie Rennie for her editorial support of this report. Finally, we would like to thank those researchers who reviewed this work, including Mary Lee and Martijn Rasser, whose feedback provided valuable refinement of the material presented in this report.

Summary

This report presents the development and implementation of a technology road-mapping process to help the Army understand the implications of key emerging technologies that could be important for Army missions in the years 2035 to 2050. Our objective is to assist the Army in preparing for shifting operational environments, including situations it might not have faced extensively in the past, such as operations facing climate change–driven extreme weather conditions. In these and other operating conditions, emerging technologies might help the Army succeed in key missions and promote American interests.

Worldbuilding for Future Contingencies

Our specific approach focused on scenario-based technology assessment. To support this scenario-based approach, exogenous drivers of the future were identified. In determining these exogenous factors, we focused specifically on identifying the most-impactful drivers outside the control of Army decisionmakers. Using expert judgment, we identified variation in the number and distribution of global powers, the trajectory of technological innovation, and the types of contingencies as the most important variables conditioning future worlds.

Variations in these exogenous factors were used to define four possible worlds and five types of conflict to develop a list of 20 illustrative scenarios that require the U.S. Army in the 2035–2050 time frame. Figure S.1 provides an illustration of the 20 scenarios across four possible future worlds.

Following the development of the 20 illustrative scenarios, we worked with Army stakeholders to down-select and prioritize specific scenarios for more fulsome development and for use in technology assessments. The five scenarios chosen were Edison Abroad, Battle of the Arctic Depths, Fog of War Machines, Thomas Schelling in the Democratic People's Republic of Korea (DPRK), and Gulf War III. Within this report, the Battle of the Arctic Depths scenario is used to demonstrate the implementation of the technology road-mapping process.

FIGURE S.1
Illustrative Scenarios Across Future Worlds

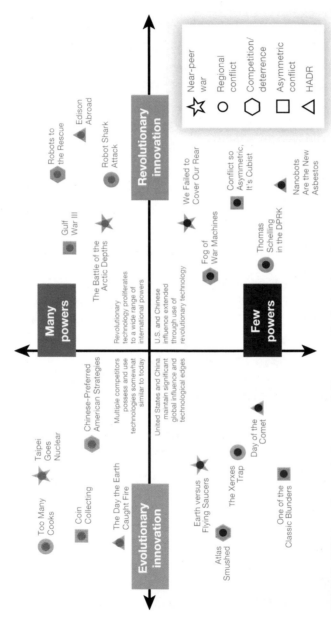

NOTES: The RAND Arroyo Center project team generated the names of the developed scenarios.
HADR = humanitarian assistance and disaster relief.

Technology Assessment for Future Contingencies

The technology road-mapping process developed in this work is a scenario-based technology assessment that builds from existing literature from Garcia and Bray.[1] The process provides structured steps to translate scenarios into mission overviews, decompose these mission steps into key challenges, identify technology candidates for the challenges, and, ultimately, assess the alignment of Army modernization priorities to candidate technology solutions and challenges. The specific steps involved in this framework are presented in Figure S.2.

The approach begins with the scenarios developed through the aforementioned process. A key component of these scenarios is a mission overview, which contains the sequential activities necessary to accomplish the mission. For each of these activities, specific challenges must be overcome to accomplish the mission, and Step 3 involves working with subject-matter experts (SMEs) to articulate these challenges in detail. Alongside this work with the scenario, we conducted an independent scan of the technology horizon. This scan produced a set of technology research areas across multiple domains (e.g., the physical domain, biological domain, and information domain). In Step 4b, we worked to align these technologies as potential solutions to the challenges identified in Step 3. In Step 5, we identify current Army modernization efforts to also align those efforts to the challenges. Finally, in Step 6, we assess the technology according to several dimensions. These include the overall readiness of the technology, interdependence with other technologies, where technology development is occurring, and the extent to which Army modernization priorities align with technology development and satisfy the identified challenges.

---

[1]   Marie L. Garcia and Olin H. Bray, *Fundamentals of Technology Roadmapping*, Albuquerque, N.M.: Strategic Business Development, Sandia National Laboratories, SAND97-0665, 1997.

FIGURE S.2

Our Scenario-Based Technology Assessment Approach

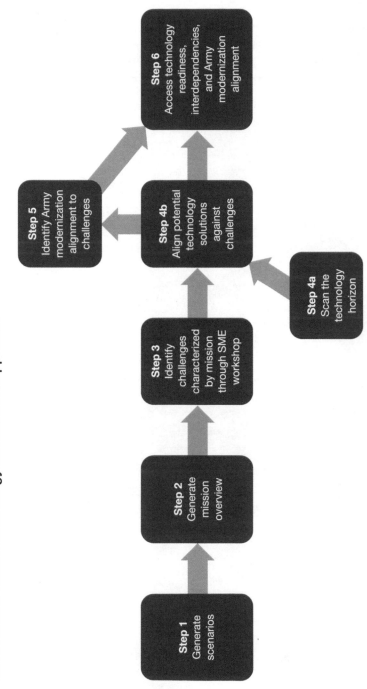

Step 1
Generate scenarios

Step 2
Generate mission overview

Step 3
Identify challenges characterized by mission through SME workshop

Step 4a
Scan the technology horizon

Step 4b
Align potential technology solutions against challenges

Step 5
Identify Army modernization alignment to challenges

Step 6
Access technology readiness, interdependencies, and Army modernization alignment

Insights and Recommendations

As demonstrated, a technology assessment process can be structured to provide a repeatable process to effectively decompose scenarios into associated challenges, identify potential candidate technology solutions to those challenges, and, ultimately, assess the alignment of Army modernization priorities against those candidate technology solutions and challenges. In conducting this research, insights emerged on both the scenario for the Arctic Depths discussed here and on the development of a technology assessment process to support such an activity.

Arctic Depths Technology Assessment Insights and Recommendations

Key Findings from the Arctic Depths

In the Battle of the Arctic Depths, the United States and its North Atlantic Treaty Organization (NATO) allies find themselves in armed conflict with a near-peer state, China, in the Arctic. The conflict is initiated under the guise of a scientific exploration trip in which Chinese forces plant a flag and claim sovereignty over a significant swath of Arctic territory claimed by Denmark. This conflict requires significant operations in an austere environment to monitor Chinese forces, expel them from the region, and provide overwatch for extended periods.

The technology assessment for the Arctic Depths enabled us to uncover insights regarding specific technologies applicable in such a scenario. For this scenario, technology areas of lubricants, quantum, space transport, air vehicles, autonomous weapon systems (AWS), and biotechnology were analyzed in detail. The following are the key insights derived from the assessment of these technologies:

- Many Army modernization priorities are aligned to the key challenges that the Army would face in the scenario. This alignment indicates that the focus and current programs within the Army might help build a foundation for the integration of the identified technology when it matures.

- – We assessed that three Army modernization priorities were aligned
  with the technology requirements in the scenario: AWS, biotechnology, and quantum technologies.
- However, some current Army modernization priorities exhibited misalignment or no alignment for the identified challenges in the Arctic Depths scenario.
  - – For lubricants, the Army is pursuing a single specification lubricant
    while development in the commercial sector is advancing lubricant
    technologies for specific operating environments.
  - – The misalignment around vertical take-off and landing (VTOL) was
    specific to the Arctic Depths scenario, which would require heavy-lift capabilities—and, given that the Army is currently focused on
    VTOL advances in the light- and medium-duty lift air assets, there
    is a potential capability gap.
- The assessment of the Arctic Depths scenario found that there was no
  alignment between Army priorities and the potential need for point-to-point space transportation.
  - – This does not necessarily mean that something is lacking in Army
    modernization; for these cases, development may be driven by commercial interests or by other services with a different mission set.
- There are unlikely to be purely technology-based solutions to many
  challenges, and technology itself potentially introduces new risks.
  Investment in technologies may have ancillary benefits and costs that
  should also be considered, which have not been considered in the
  exploration of this Arctic Depths scenario.

Recommendations from the Arctic Depths

In conducting the technology assessment for the Arctic Depths scenario,
specific cases of misalignment of Army modernization priorities and
technology development were identified. In these instances, the Army
Futures Command (AFC) may consider communicating the following
recommendations:

- The Army, likely through Headquarters Department of the Army and
  Army Materiel Command, should develop key partnerships in the pri-

vate sector (e.g., with industry leaders in energy, space transport, and the like) and with multinational partners (e.g., key allies within NATO) to ensure the availability and integration of technology it needs, especially for the technologies identified as critical needs through technology assessment. These partnerships may be supported through the establishment and funding of centers of excellence.

- Where misalignment in technology development and Army modernization priorities exist, the Headquarters Department of the Army, Army Materiel Command, and AFC should consider investing resources, updating doctrine, and implementing alternative strategies to take advantage of the technologies identified as critical needs.
- To provide a more balanced perspective if comparisons across scenario-based technology assessments are desired, AFC should further develop this portfolio of scenarios to include scenarios where technological advancements are evolutionary in nature. This step would allow Army stakeholders to provide a broader perspective from technology assessment across the exogenous uncertainty of technology development.

Technology Assessment Process Insights and Recommendations

Key Findings on Technology Assessment

In addition to deriving insights on specific technologies, this work also provided a lens to view findings on the development of technology assessments. In developing the technology road-mapping process described, the study team provides the following insights:

- Many technology assessment concepts, such as technology road-mapping, constructive technology assessment, strategic technology assessment, and backcasting, could be used for assessment activities based on the type of information available and desired outcome of the activity.
  - Selection of the appropriate concept to employ should be based on what the information generated during a technology assessment will be used to support.

- For the work conducted here, the desired end state of the technology assessment was to provide a forward-looking analysis on how technology may be applied to future scenarios, which necessitated the use of technology road-mapping.
  - Pursuing technology road-mapping under this effort required the development of scenarios to provide a specific future against which technologies could be evaluated.
- Applying the technology road-mapping process for these scenarios necessitated consideration of technologies with high degrees of uncertainty. This deep uncertainty about the future required the use of SMEs to generate information about technology readiness, impacts, and utility for the scenario studied.
  - The use of SMEs to solicit this information may be required in instances of deep uncertainty, and care should be taken in identifying appropriate expertise.
- A systematic process should be in place to provide a repeatable process for the assessment of technologies against future scenarios. The developed technology road-mapping process detailed in this report provides (1) a framework for technology road-mapping activities and (2) the structure to repeatably decompose scenarios into sets of challenges, align technologies to challenges, and assess Army modernization priorities against candidate technology solutions.

Recommendations for Future Technology Assessment

To operationalize these findings on technology assessments, the Army should consider implementing technology assessment frameworks to provide structured, repeatable processes for future activities. To support this effort, the Army should consider the following recommendations:

- The Army should be prepared for a broad range of future contingencies and world states where technology is diffuse, and the United States does not have a technological advantage. To support this, the Army ought to further develop the portfolio of scenarios considered under technology assessment activities.

- The Army should consider the desired outcomes for future technology assessments to ensure that the appropriate technology concept is applied.
- For future technology road-mapping activities reliant on SME input, the Army should implement a structured process based on best practices to elicit information, such as that provided in Appendix B.
- The Army should leverage technology assessment methods to address modernization priorities and requirement gaps. In conducting further technology assessments, the Army should
  - identify techniques to determine the robustness of candidate technology solutions across scenarios
  - consider soliciting information from alternative sources, such as crowd-forecasting
  - continually re-evaluate the current state of knowledge for future technologies and consider methods integrating quantitative modeling and simulation when sufficient certainty exists to support such activities.

The future is uncertain, and predictions of technology are unlikely to be true. That said, conducting technology assessments in rigorous and robust ways might enable the Army to best prepare itself for future contingencies and operating environments.

Contents

About This Report iii

Summary v

Figures and Tables xvii

CHAPTER ONE

Introduction 1

    Objectives of This Research 1

    Challenges and Opportunities of Technology Forecasting................. 2

    Our Approach 4

CHAPTER TWO

Worldbuilding for Future Contingencies 7

    Drivers of the Future 7

    Developing Illustrative Scenarios... 14

CHAPTER THREE

Technology Assessments for Future Contingencies 19

    Background on Technology Assessments.................................. 19

    Scenario-Based Technology Assessment Approach 25

    Technology Assessment for the Battle of the Arctic Depths Scenario 28

CHAPTER FOUR

Insights and Recommendations 49

    Arctic Depths Technology Assessment Insights and

        Recommendations... 50

    Technology Assessment Process Insights and Recommendations 54

APPENDIXES

A. Prioritized Scenario Descriptions... 57

B. Sample Assessment Activity: Technology Matching 69

C. Battle of the Arctic Depths Technology Research 75

Abbreviations... 89

References 91

Figures and Tables

Figures

S.1. Illustrative Scenarios Across Future Worlds....................... vi
S.2. Our Scenario-Based Technology Assessment Approach....... viii
2.1. Illustrative Scenarios Across Future Worlds...................... 15
3.1. Decision Tree to Help Identify Type of Technology
Assessment Concept to Use 23
3.2. Our Scenario-Based Technology Assessment Approach......... 26
B.1. Technology Domains... 72

Tables

2.1. Key Drivers of the Future Varied to Develop Future
Scenarios.. 9
3.1. Snapshot of Battle of the Arctic Depths Strategies-to-Tasks
Challenge Identification 34
3.2. Battle of the Arctic Depths Technology Alignment 36
3.3. Battle of the Arctic Depths Army Modernization Alignment... 39
3.4. Battle of the Arctic Depths Technology and Army
Modernization Assessment.. 42
C.1. Cold Weather Properties of Base Oils 76
C.2. Cold Weather Properties of Greases 77

CHAPTER ONE

Introduction

This research seeks to develop a process to help the Army understand the implications of key emerging technologies that may be important for Army missions in the years 2035 to 2050. There are many possible ways of evaluating and forecasting the trajectory of emerging technology. In this report, we present a specific approach focused on scenario-based technology assessment. In this chapter, we explain the goals of our research, clarify the limitations and possibilities of technology forecasting, and provide an overview of the approach that will be further developed in subsequent chapters.

Objectives of This Research

The future is highly uncertain; that said, it is important for the Army to try to anticipate future global developments and technological changes. The objective of this forecasting is to assist the Army in preparing for shifting operational environments, including environments it might not have faced extensively in the past, such as in extreme weather conditions driven by climate change. In these and other operating conditions, emerging technologies might help the Army succeed in key missions and promote U.S. interests. Forecasting could also help the Army better understand and anticipate the types of conflicts it might face, along with the characteristics of key adversaries, and the operational-level challenges that could be in play. Preparing and planning for future contingencies are especially important in the context of scarce resources in an austere budget environment where the Army already today needs to make difficult decisions about how it dedicates its resources.

In addition, forecasting can help the Army build the necessary partnerships it needs for the development, acquisition, and integration of emerging technologies. Of course, these partnerships include relationships across the U.S. Department of Defense (DoD) enterprise to ensure that the Army is working jointly and able to leverage the technologies that other services may have available. Key partnerships also include the private industry organizations and researchers that often lead the development of new technologies. By anticipating future challenges and technological opportunities, the Army might also more effectively work across the U.S. interagency and with Congress to ensure that its funding is sufficient and that resources are dedicated to important technologies. Lastly, technological forecasting could help the Army lay the groundwork with international allies and partners to ensure that the United States can integrate emerging technologies in multinational operations, and that comparative advantages across these partners are used effectively.

The goal of this research is to develop and execute a process that will help identify potential emerging technology trends beyond 2035 and describe the military implications of these technologies. We hope to assist the Army in developing tools that can provide insight into the military contingencies in which it might be required to engage and the implications of new technologies for the Army to meet future operational challenges. In doing so, it seeks to lay the groundwork for technology forecasting processes that help the Army be prepared for future environments, adversaries, and challenges such that the Army can take the necessary steps today to meet those future requirements.

Challenges and Opportunities of Technology Forecasting

In the introduction to her classic 1969 novel, *The Left Hand of Darkness*, science fiction author Ursula K. Le Guin observed that "prediction is the business of prophets, clairvoyants, and futurologists. It is not the business of novelists. A novelist's business is lying." She continued that "the weather bureau will tell you what next Tuesday will be like, and the Rand Corporation [sic] will tell you what the twenty-first century will be like. I don't recommend that you turn to the writers of fiction for such information. It's none of their business."[1]

With the benefit of hindsight, however, the predictions of futurologists of the past about what our current century would be like seem no more prophetic than the writings of science fiction writers, such as Le Guin. In a 1964 RAND study based on a Delphi survey of expert opinion, RAND researchers made concrete forecasts for the then-distant year 2000, when fusion power plants would be "a new source of energy" and militaries would manipulate the weather to their advantage. Meanwhile, "on the Moon, mining and manufacture of propellant materials will be in progress," and a manned mission would have reached Mars. To their credit, the study's authors qualified that "No claims are made, or can be made, for the reliability of the predictions obtained here," but, even in 2021, these prognostications still remain in the realm of science fiction.[2]

Predicting the future is hard—even for professionals. But as the authors of the 1964 RAND study explained, such predictions are necessary even though they are seldom accurate. The authors note that because

> trend predictions—implicit or explicit, "scientific" or intuitive—about periods as far as twenty or even fifty years in the future do affect current planning decisions (or lack of same) . . . almost anything further we can learn about the basis, the accuracy, and the means for improving such long-term forecasts will be of value.[3]

Therefore, well-reasoned predictions may have value even if they are unlikely to be accurate. The report's introduction further explained that "inasmuch as they reflect explicit, reasoned, self-aware opinions, expressed in light of the opinions of associate experts," predictions such as those made in the report "should lessen the chance of surprise and provide a sounder

---

[1] Ursula K. Le Guin, *Hainish Novels and Stories*, Vol. 1, New York: Library of America, 2017, p. 1023.

[2] Theodore J. Gordon and Olaf Helmer-Hirschberg, *Report on a Long-Range Forecasting Study*, Santa Monica, Calif.: RAND Corporation, P-2982, 1964, p. vi.

[3] Gordon and Helmer-Hirschberg, 1964, p. v.

basis for long-range decisionmaking than do purely implicit, unarticulated, intuitive judgments."[4]

The future is fraught with uncertainties, both geostrategic and technological. However, the intuitions of such experts as those interviewed for the 1964 RAND forecasting study may not be meaningfully more accurate than those of laypeople. Another source of uncertainty is that actual, observed outcomes were not necessarily likely to occur; improbable events regularly occur over the course of history. Nor can such events be expected to cancel each other out in a nonlinear system; instead, they can compound with each other and produce outcomes that might have been deemed especially improbable by well-regarded experts. Therefore, even the most-rigorous and thoughtful forecasts (in the sense that they correctly predict the likelihood of possible outcomes on the basis of all available evidence) cannot be counted on to be predictive.

There is another way, however, that predictions of the future do not need to be actually predictive to be useful. In many cases, predictions are made in the hopes that they will become antipredictive by becoming self-negating prophecies. The objective may try to foreclose the possibility of certain futures that may not seem especially likely but appear so terrible that we need to take steps to prevent them anyway. Predictions about possible threats from new military technologies are perhaps likely to take this form. For instance, there may be a possibility that an adversary could be emboldened by its sudden acquisition of an unprecedented military capability so as to engage in aggression when the adversary otherwise would not have. If we can anticipate this capability and develop plausible countermeasures before the adversary matures it, the adversary may decide not to field the capability at all. It is our hope that the technology assessment process we describe can help identify a few wrong predictions of this kind.

Our Approach

There are many ways to conduct technological forecasting and assessment (see Chapter Three for some specific options). The approach we have taken

---

[4] Gordon and Helmer-Hirschberg, 1964, p. vi.

is a scenario-based technological assessment that leverages the views of experts following a sequenced probe.

The first step for this approach is to develop a set of possible future scenarios. In Chapter Two, we lay out how we developed illustrative scenarios depicting military contingencies requiring the Army in the 2035–2050 time frame. Our approach begins with identifying the key drivers of the future that are outside the control of Army decisionmakers, but are nonetheless important in determining the long-term success of Army strategies. With these key drivers in mind, the research team developed a set of four possible future worlds, each of which set a global context for five types of potential military contingencies requiring Army engagement. We used this set of four worlds and five possible contingencies in each world to develop 20 operational scenarios.

The technology assessment process we employed is an involved process, so it was not possible to do technology assessments for all 20 scenarios. Instead, we worked with Army Futures Command (AFC) stakeholders to prioritize scenarios for the full assessment. In the context of this technology assessment, we also looked closely at current Army modernization priorities, including those aligned with the Army's Cross-Functional Teams (CFTs). To best inform Army planning, we assessed the extent to which existing Army modernization priorities align with the operational requirements articulated in the scenarios. This enabled concrete recommendations designed to help shape Army decisions about its modernization priorities.

In Chapter Three, we present different approaches to technology assessment and how we performed the assessment in a specific scenario focused on an illustrative conflict with China in the Arctic in the year 2045. In Chapter Four, we present our findings and associated recommendations on the role and approach for conducting technology assessment, along with insights and recommendations related to specific emerging technologies.

CHAPTER TWO

Worldbuilding for Future Contingencies

This chapter lays out the process we employed to develop 20 scenarios that occur in the 2035–2050 time frame depicting contingencies requiring the Army.

Drivers of the Future

The long-term future is deeply uncertain, and decisionmaking for an indeterminant future might ideally require that decisions be robust to an array of plausible futures.[1] Although no single method provides a panacea, numerous techniques exist to facilitate planning for the future, including modeling and simulation, wargaming, and scenario generation.

We have used an approach that seeks to develop possible future worlds and specific scenarios to structure development of long-term analysis that could be used to facilitate technology assessment and to evaluate the Army's modernization strategy.[2]

We began with identifying a set of exogenous factors that are the key drivers of future worlds. The key drivers we sought to identify were those we

---

[1] Robert J. Lempert, Steven W. Popper, and Steven C. Bankes, *Shaping the Next One Hundred Years: New Methods for Quantitative, Long-Term Policy Analysis*, Santa Monica, Calif.: RAND Corporation, MR-1626-RPC, 2003.

[2] We have based our approach on an approach laid out in Peter Schwartz, *The Art of the Long View: Planning for the Future in an Uncertain World*, New York City: Penguin Random House, 1996; Lempert, Popper, and Bankes, 2003.

deemed most relevant for conditioning future operating environments and adversaries, the types of challenges the Army might face, and the sorts of future technologies with military implications. In our approach, the most-important drivers of the future are those outside the control of the Army decisionmakers, which may nonetheless prove important in determining the success of their strategies. With these key drivers in mind, we could evaluate the near-term actions that comprise the strategies Army decision-makers want to evaluate. These actions include such things as Army modernization priorities and decisions regarding technology investment and integration. Finally, the process also involves an element that captures the standards that decisionmakers and other interested communities would use to rank the desirability of various decisions.[3]

This research is designed to help the Army understand the technologies that will be important for future contingencies and to analyze whether Army modernization priorities satisfy the identified technology requirements. Given this objective, the structure becomes exceedingly complex, with interactions between the exogenous factors, policy actions, relationships between the factors and choices, and the measures used to evaluate outcomes. In this study, we chose to assume that the exogenous factors were truly exogenous—that is, changes in the levers, relationships, and measures do not feed back into the exogenous factors. This allowed us to (1) identify the most important exogenous factors and (2) generate scenarios that are illustrative across future worlds.

To determine the exogenous factors, we focused specifically on identifying the most impactful factors relevant to defining a range of potential worlds that are outside the control of Army decisionmakers. Using expert judgement, we identified variation in the number and distribution of global powers, the trajectory of technological innovation, and the types of contingencies as the most important variables conditioning future worlds.[4]

---

[3] Lempert, Popper, and Bankes, 2003.

[4] Our identification of factors aligns closely with AFC's own identification of factors presented in AFC, *Future Operational Environment: Forging the Future in an Uncertain World, 2035–2050*, Austin, Tex.: U.S. Army Futures Command, AFC PAM 525-2, 2019. However, we came to these factors independently and before we were aware of AFC's work.

Although there are some things the Army might do to shape the outcome of these factors, we believe that there are many aspects that will be simply outside the control of Army decisionmakers, including decisions by civilian leadership, actions by external global forces, and the characteristics of specific technologies and the industry and marketplace surrounding it. In the 2035–2050 time frame, these three factors are uncertain, could fluctuate, and yet will likely significantly affect what missions and objectives are tasked to the Army. Therefore, varying these factors to develop illustrative scenarios enables making informed conjectures about the extent that the U.S. Army's modernization priorities satisfy the operational challenges identified and can help explore how these decisions are robust to a variety of potential futures.

Key Exogenous Drivers of the Future

We identified three exogenous drivers that were varied to develop sets of future worlds and illustrative scenarios. Table 2.1 summarizes these three exogenous drivers varied in this study.

First, the global powers factor illuminates the implications of variation in the number of global powers that the United States will confront in the 2035–2050 period. We define global powers as those actors that can exert significant global influence to achieve political, economic, or other objectives. We treated this factor as binary and its values could either be "few" or "many." In future worlds where the United States confronts few global powers, the few global powers include current near peers (i.e., China and Russia). In future worlds where the United States confronts many global powers, such worlds include additional state actors (such as Turkey, Brazil,

TABLE 2.1

Key Drivers of the Future Varied to Develop Future Scenarios

Key Exogenous Drivers	Possible Values
Number of global powers	Few or many
Technology trajectory	Evolutionary or revolutionary
Types of contingencies	Competition and deterrence; near-peer conflict; regional conflict; asymmetric conflict; humanitarian assistance and disaster relief (HADR)

India, and others) and nonstate actors (such as multinational corporations or international terrorist organizations) that can substantially exert global influence.

Second, the innovation trajectory variable attempts to elucidate how variation in the maturation and military diffusion of innovative, emerging technologies might influence the nature of conflict. The technology innovation trajectory consists of the life cycle of emerging technologies, from initial research and development (R&D) to maturity and integration in commercial and military applications. The trajectory of technology innovation depends on R&D decisions, investment from public and private sector actors, legal and political enablers (e.g., supportive regulatory environment), military budget and decisionmaking, access to skilled personnel, and other human factors.

Technologies will have different trajectories, and some trajectories might be interdependent with the development of other technologies, while others are independent. However, for the purposes of tractability, we have simplified this variable into a binary one, and its values are either *evolutionary* or *revolutionary*. For the purposes of this study, we define *evolutionary technology change* as a slow or linear path to maturity in military applications. However, we define *revolutionary technology change* as a rapid or exponential path to maturity in military applications. In consultation with Army stakeholders, we prioritized scenarios within the revolutionary technology innovation worlds, assuming that if the Army's modernization priorities were robust to such an outcome, they might also be robust to scenarios where less innovation occurred.

Finally, although the coming decades might seem poised to experience a return to great power competition, it is plausible that the Army will be asked to operate in a variety of different types of military contingencies.[5] The Cold War produced a "Long Peace" only as it related to direct great-power conflict; large, violent struggles were nonetheless fought to determine the

---

[5] James Mattis, *Summary of the 2018 National Defense Strategy of the United States of America: Sharpening the American Military's Competitive Edge*, Washington, D.C.: U.S. Department of Defense, 2018.

terms of the world order.[6] Thus, the United States should expect to confront a range of deterrence mission, competition requirements, contingencies, and other demands in the 2035–2050 time frame, including competition and deterrence activities, full-on near-peer conflict, regional conflict with regional powers, asymmetric conflict with nonstate actors, and HADR missions. We classify *competition and deterrence* as military activity to reduce military power or nonmilitary influence of a potential adversary nation-state. *Near-peer conflict* is seen as armed conflict with a state that roughly matches the military capabilities brought to bear by the United States (i.e., China and Russia). *Regional conflict* is classified as armed conflict with a state with military capabilities that allow it dominance in the region, but which do not rival that of the United States (i.e., Democratic People's Republic of Korea [DPRK] and Iran). *Asymmetric activity* refers to those contingencies involving such nonstate actors as an insurgent groups or terrorists. Finally, *HADR contingencies* encompass circumstances in which military activities are leveraged to reduce suffering of civilians that results from causes other than armed conflict or terrorism.

Additional Assumptions About the Future

Beyond the three exogenous factors discussed above that were varied in this study to produce future worlds, there are many other potential exogenous factors that might condition the future. Any number of developments have the potential to result in meaningful variation in the future security environment that the United States might confront. To bound this space, and thus the number of futures considered, the authors made assumptions about *nondefense–related* global trends. These assumptions were held constant across all futures rather than varied. Specifically, we assumed that the changing global climate, the transition to alternative sources of energy, population growth, and a contested information environment, would be relatively inelastic across various futures.

---

[6] John Lewis Gaddis, *The Long Peace: Inquiries into the History of the Cold War*, New York: Oxford University Press, 1987; and Paul Thomas Chamberlin, *The Cold War's Killing Fields: Rethinking the Long Peace*, New York: Harper Collins, 2019.

We assumed that the climate and environment continue to change, causing climate-related instability.[7] Humanity does not exist separate from the biosphere. Frequent and unpredictable natural disasters, the increased spread of vector-borne infectious diseases, and resource scarcity will lead to collapsed economies, failed states, and widespread climate migration.[8] Population trends establish underlying conditions that shape and catalyze events in the political and economic spheres.[9] Motivated by the social and economic effects of climate change, we assumed that the global economy will continue to transition away from fossil fuels and toward alternative sources of energy.

Trends suggest that the global population will reach 8 billion to 10.5 billion between the years 2040 and 2050. Population trends by themselves are neither inherently positive nor negative. Rather, populations affect national security indirectly, interacting with the political capacity of the state. If the state possesses legitimate and effective institutions, governance, infrastructure—i.e., political capacity—population growth can spur economic growth. Absent economic and social prospects enabled by this capacity, though, population growth can instead spur political instability or conflict.[10] Projections suggest that the ten most-populous countries in 2050 will be India, China, Nigeria, the United States, Bangladesh, Indonesia, Pakistan, Brazil, the Democratic Republic of Congo, and Ethiopia.[11] Other than the United States, at present, these countries lack political capacity, and they are relatively poor and conflict-prone. If current trends continue, this pop-

---

[7]   C. E. Richards, R. C. Lupton, and J. M. Allwood, "Re-framing the Threat of Global Warming: An Empirical Causal Loop Diagram of Climate Change, Food Insecurity and Societal Collapse," *Climatic Change*, Vol. 164, No. 49, 2021.

[8]   Emrah Sofuoğlu and Ahmet Ay, "The Relationship Between Climate Change and Political Instability: The Case of MENA Countries (1985:01–2016:12)," *Environmental Science and Pollution Research*, Vol. 27, April 2020; Huei-Ting Tsai and Tzu-Ming Liu, "Effects of Global Climate Change on Disease Epidemics and Social Instability Around the World," Asker, Norway, Human Security and Climate Change International Workshop, June 21–23, 2005.

[9]   Jennifer Dabbs Sciubba, "Demography and Instability in the Developing World," *Orbis*, Vol. 56, No. 2, 2012.

[10]   Sofuoğlu and Ahmet, 2020, p. 14033; Sciubba, 2012, p. 269.

[11]   United Nations, "2019 Revision of World Population Prospects," webpage, undated.

ulation growth will further increase the number of mega-cities, multiplying environmental strains and population-resource imbalances, which can foment civil conflict.[12]

Information and perception–manipulation technologies will become even more pervasive, leading to the information environment being regularly contested. These kinds of technologies will allow many actors—individuals, businesses, and governments—to conduct sophisticated influence operations via deepfakes, microtargeting, machine learning–driven programs, and spoofing algorithms.[13] In addition to the economic impacts of information technology, social trends interact with these technologies.[14] Information and perception manipulation technologies will continue to reshape the relationship between the state and society, with potentially grave consequences. As Copeland noted in 2000, "the declining sovereignty and legitimacy of the nation-state as it struggles to respond to economic, social, and political challenges is brought on by the information revolution."[15]

We also made assumptions about defense-related trends in the United States from a macro perspective. We assumed the United States will remain engaged in the international system and will retain the ability to exert influence and project power on a global scale. The United States will confront a range of demands but be constrained by insufficient resources to accomplish all goals and take on all challenges. Therefore, the United States will have to prioritize regions and adversaries, but we retained the current emphasis in the National Defense Strategy that identifies China, Russia, the

---

[12] Sciubba, 2012, p. 275.

[13] Michael J. Mazarr, Ashley L. Rhoades, Nathan Beauchamp-Mustafaga, Alexis A. Blanc, Derek Eaton, Katie Feistel, Edward Geist, Timothy R. Heath, Christian Johnson, Krista Langeland, Jasmin Léveillé, Dara Massicot, Samantha McBirney, Stephanie Pezard, Clint Reach, Padmaja Vedula, Emily Yoder, *Disrupting Deterrence: Examining the Effects of Technologies on Strategic Deterrence in the 21st Century*, Santa Monica, Calif.: RAND Corporation, RR-A595-1, 2022, p. 30.

[14] Manuel Castells, *The Information Age: Economy, Society and Culture*, Vol. 1: *The Rise of the Network Society*, Hoboken, N.J.: Wiley-Blackwell, 1996.

[15] Thomas E. Copeland, ed., *The Information Revolution and National Security*, Carlisle Barracks, Pa.: U.S. Army War College, August 2000, p. 1.

DPRK, and Iran as the principal priorities.[16] Yet, we tried to create scenarios that emphasized the potentially unexpected ways in which the United States could encounter these competitors.

Developing Illustrative Scenarios

One of the most prolific and influential early advocates of scenarios as a tool of analysis, the nuclear strategist and futurist Herman Kahn defined scenarios as "attempts to describe in some detail a hypothetical sequence of events that could lead plausibly to the situation envisaged." Kahn listed several advantages of scenarios, including that "[t]hey serve to call attention, sometimes dramatically and persuasively, to the larger range of possibilities that must be considered in the analysis of the future," that "[t]hey force the analyst to deal with details and dynamics that he might easily avoid treating if he restricted himself to abstract considerations," and that "[t]hey can illustrate forcefully, sometimes in oversimplified fashion, certain principles, issues, or questions that might be ignored or lost if one insisted on taking examples only from the complex and controversial real world." He also acknowledged several criticisms, not the least of which was "that scenarios may be so divorced from reality as not only to be useless but also misleading, and therefore dangerous." But in Kahn's view, realism was not necessarily the point of scenarios, as "one must remember that the scenario is not used as a predictive device" because "the analyst is dealing with the unknown and to some degree unknowable future." Kahn emphasized that "imagination has always been one of the principal means for dealing in various ways with the future, and the scenario is simply one of many devices useful in stimulating and disciplining the imagination."[17]

We used the four possible worlds and five types of conflict to develop a list of 20 illustrative scenarios that require the Army in the 2035–2050 time frame. Figure 2.1 plots this list of 20 scenarios across four possible future worlds.

---

[16] Mattis, 2018.

[17] Herman Kahn and Anthony J. Weiner, *The Year 2000: A Framework for Speculation on the Next Thirty-Three Years*, New York: Macmillan Publishers, 1967.

FIGURE 2.1
Illustrative Scenarios Across Future Worlds

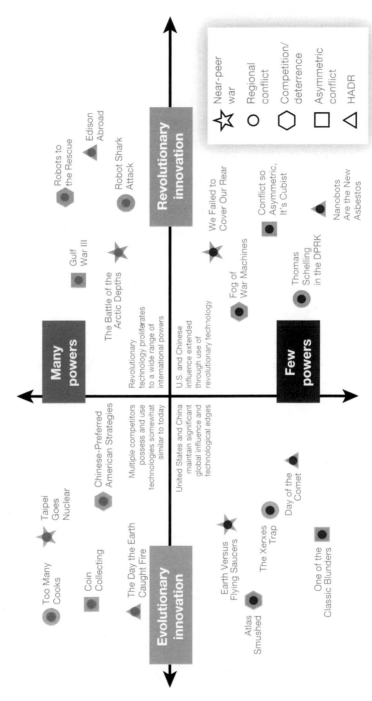

NOTES: The RAND Arroyo Center project team generated the names of the developed scenarios.

The initial 20 scenarios resulted from a process intended to produce a cross section of geostrategic situations, rates of technological progress, and mission types. The goal was to briefly outline at least one scenario with every permutation of the considered possibilities in each of these three categories. After consultation with AFC stakeholders, we then selected a subset of these 20 concise scenarios to be fleshed out into detailed scenarios that could serve as the basis for in-depth analyses.

The two geostrategic situations employed for scenario development hinged upon the role of the United States in global affairs. In half of the scenarios, the United States remained at the center of the international order, whether through international consent or the weakness of prospective rivals. Rogue and rival powers still existed in most of these scenarios, but these were not in the position to mount a serious challenge to displace the United States as top power. In the other half of the scenarios, the United States had ceased to be the leading power—however, the United States being supplanted by a rising state, such as China, was the cause only in some of these cases. In the other cases, a multipolar order emerged because the United States had become isolationist, or because shifts in relative power increased the ability of less influential states to pursue independent courses of actions against the wishes of the largest powers. It should be emphasized that this category was about relative rather than absolute power: A stagnant United States could still remain the top power if other states were weakening, while a swiftly advancing United States could still be surpassed if a rival power were advancing even more quickly.

Similarly, two broad categories served to represent different rates of technological progress. In the first of these, worlds with revolutionary technology, technology advances rapidly across a broad array of application areas; in the latter worlds with evolutionary technology, technological advance is slower and/or restricted to a limited number of applications. These two categories are not meant to imply that all kinds of technological progress will proceed in lockstep, but to define different kinds of technological landscapes. Evolutionary technology worlds could include isolated examples of rapid technological advancement, while revolutionary worlds might include certain areas in which technology remained stagnant. In the most extreme evolutionary scenarios, technology could actually regress from present-day levels (for example, because the Earth was struck by a comet).

Between the two possible geostrategic futures and the two technological environments there resulted four possible contexts in which scenarios might take place (U.S.-centric revolutionary technology, U.S.-centric evolutionary technology, multipolar revolutionary technology, and multipolar evolutionary technology).

These were then combined with five different mission types to define 20 different scenario prompts, each of which consisted of a context and one of the missions.

The objective of these scenarios was not to try to predict the likeliest scenario for a mission of the chosen type in the stipulated world. Instead, the goal was to illuminate particularly compelling or tricky challenges to stimulate later analysis. We sought to create scenarios that would stimulate the imaginations of others contemplating them, and one way to do this was to envision unfamiliar circumstances to shock those considering the scenarios into unconventional thinking. We should not limit ourselves to only imagining futures that are basically like today, albeit with some additional technology. If anything, geostrategic surprises—such as the sudden nonviolent collapse of the Soviet bloc or the September 11, 2001, terrorist attacks—might be more common in practice than technological surprises. Moreover, our intuitions about what is likely, or even possible, may not be very accurate, as the predictions made by the 1964 RAND futurological study show. Therefore, when it comes to scenario development, wilder is arguably better. Exotic scenarios better stretch the imagination and jolt analysts out of their comfortable assumptions, and these scenarios may be no less informative than staid, conventional ones. Taking these observations to heart, this study postulated a variety of scenarios ranging from the commonsensical (e.g., future reiterations of conflicts that have already occurred) to extremely fanciful (e.g., a world in which the U.S. military comes under sustained attack by unidentified aerial phenomena of possible nonterrestrial origin).

Prioritizing Scenarios

Following the development of the 20 illustrative scenarios, the research team worked with Army stakeholders to down select specific scenarios to be prioritized for more fulsome development and to be used for technology assessments. The five scenarios chosen were Edison Abroad, Battle of the

Arctic Depths, Fog of War Machines, Thomas Schelling in the DPRK, and Gulf War III. We developed detailed slides presenting the environmental conditions, adversary characteristics, mission steps associated with these scenarios based on a template and set of examples provided by AFC. These detailed scenario slides are an independent product from this research to be used for Army wargaming, technology assessments, or other activities.[18] Appendix A includes overviews of these scenarios.

---

[18] These slides were provided to the Army for use in tabletop exercises and therefore are not available to the general public.

CHAPTER THREE

Technology Assessments for Future Contingencies

This chapter provides background on technology assessments approaches, presents the specific approach to scenario-based technology assessment we employed, and explores the implications of key technologies in a specific scenario, the Battle of the Arctic Depths, as an example of the approach in practice.

Background on Technology Assessments

Before describing the technology approach that we used, we provide a brief overview of different approaches. Although we pursued a scenario-based approach, there are other approaches that the Army might also consider in future technology assessment activities.

Technology assessment is a term indicating a group of methods to investigate technology. Grunwald defines these as "systematic methods used to scientifically investigate the conditions for and the consequences of technology . . . and to denote their societal evaluation."[1] Banta suggests these are forms "of policy research that examines short- and long-term consequences (e.g., societal, economic, ethical, legal) of the application of technology."[2] Van Den Ender et al. suggest this is a way to "systematically appraise the

---

[1] Armin Grunwald, "Technology Assessment: Concepts and Methods," in Anthonie Meijers, ed., *The Handbook of the Philosophy of Science, Philosophy of Technology and Engineering Sciences*, Amsterdam, Netherlands: North-Holland, 2009.

[2] David Banta, "What Is Technology Assessment?" *International Journal of Technology Assessment in Health Care*, Vol. 25, No. S1, July 2009.

nature, significance, status, and merit of a technological program."[3] Because the term technology assessment encompasses a group of different concepts, various different ways can accomplish any one technology assessment, and the methods chosen can vary as a function of the context under which the research is conducted.[4]

In absence of a regulatory or similar reason to use a particular approach, an analyst can first consider the goal of the research. This can help determine the type of technology assessment concept to use. For example, some different types of technology assessments are

- **awareness technology assessment:**[5] Sometimes referred to as an *early warning assessment*, this type seeks to "detect, control, and direct technological changes and developments so as to maximize the public good while minimizing the public risks."[6] Here the future is unspecified, there is more of an interest in policy, and the research is targeted for a public good.
- **strategic technology assessment:**[7] This type is a process "consisting of analyses of technological developments and their consequences and a debate on the bases of these analysis. The technology assessment should provide information that could help the actors involved in developing their strategies and that might define subjects for fur-

---

[3] Jan Van Den Ende, Karel Mulder, Marjolijn Knot, Ellen Moors, and Philip Vergragt, "Traditional and Modern Technology Assessment: Toward a Toolkit," *Technological Forecasting and Social Change*, Vol. 58, No. 1–2, May–June 1998.

[4] Josée C. M. Van Eijndhoven, "Technology Assessment: Product or Process?" *Technological Forecasting and Social Change*, Vol. 54, No. 2–3, February–March 1997.

[5] Van Den Ende et al., 1998; R. Smits and J. Leyten, T*echnology Assessment: Waakhond of Speurhond (Technology Assessment: Watchdog or Tracker? Towards an Integral Technology Policy)*, Vrije Universiteit Amsterdam, Kerckebosch, Zeist, Vol. 28, 1991.

[6] Van Den Ende et al., 1998, pp. 6–7.

[7] Charles W. Thomas, "Strategic Technology Assessment, Future Products and Competitive Advantage," *International Journal of Technology Management*, Vol. 11, No. 5-6, January 1996; Goel Kahen, "Strategic Development, Technology Transfer and Strategic Technology Assessment in Changing Environments," *Proceeding of the First International Conference on Dynamics of Strategy*, Guildford, United Kingdom, April 11–12, 1996.

ther [technology assessment] analysis."[8] They often involve studies and activities to support a specific actor or a specific group of actors in the development of their technology policy or technology strategy. Here the future is unspecified, there is more of an interest in policy, and the research is targeted for a specific actor.

- **backcasting technology assessment:**[9] Here the future is specified, there is more of an interest looking backward in time to determine how to get to that particular future. This often includes constructing desirable futures or scenarios, then determining how to get here.

- **constructive technology assessment:**[10] Here the future is unspecified, there is more of an interest in technology, such as improving design, development, and other processes needed to increase technology readiness levels. This means that this is more of an open-ended world building exercise, which should include stakeholder elicitation methods that draw on various perspectives across disciplines.

- **technology road-mapping:**[11] This is a method that specifies the desired future and takes steps from now to the future to determine how to get there. Garcia and Bray lay out a process for this, which involves identifying the scope, boundaries, and target; identifying the critical system requirements; specifying major technology areas; specifying

---

[8] Van Den Ende et al., 1998, p. 8.

[9] Van Den Ende et al., 1998; Jaco Quist and Philip Vergragt, "Past and Future of Backcasting: The Shift to Stakeholder Participation and a Proposal for a Methodological Framework," *Futures*, Vol. 38, No. 9, 2006.

[10] Johan Schot and Arie Rip, "The Past and Future of Constructive Technology Assessment," *Technological Forecasting and Social Change*, Vol. 54, No. 2–3, 1997; Audley Genus, "Rethinking Constructive Technology Assessment as Democratic, Reflective, Discourse," *Technological Forecasting and Social Change*, Vol. 73, No. 1, 2006; Kornelia Konrad, Arie Rip, and Verena Schulze Greiving, "Constructive Technology Assessment—STS for and with Technology Actors," *European Association for the Study of Science and Technology*, Vol. 36, No. 3, 2017; and Encyclopedia.com, "Constructive Technology Assessment," webpage, undated.

[11] Marie L. Garcia, and Olin H. Bray, *Fundamentals of Technology Roadmapping*, Albuquerque, N.M.: Strategic Business Development, Sandia National Laboratories, SAND97-0665, 1997.

drivers within the technology areas; and identifying alternatives and timelines.[12]

Figure 3.1 relates these different types of technology assessments via a decision tree. For example, if the future is specified, a researcher likely wants to choose a technology road-mapping or a backcasting technology assessment; the choice is a function of whether the researcher wants to plan from now to the future (technology road-mapping) or from the future to now (backcasting technology assessment). Next, consider whether the future is unspecified. In this instance, if a researcher is more interested in developing technology, the researcher might use a constructive technology assessment. If the researcher is more interested in informing policy, then the researcher might use an awareness technology assessment (if more interested in public good) or a strategic technology assessment (if more interested in a specific actor).

Given the complexities involved with policy and different time frames of consequences, there are a large number of nonuniform methods that any one given technology assessment could use.[13] Approaches included in a technology assessment could include economic analysis, decision analysis, impact analysis, landscape analysis, market analysis, risk assessment, scenario analysis, technological forecasting, or technical performance assessment.[14] Across each of these, there are a variety of different types of methods that could be used. For example, a cost benefit analysis is certainly within the economic analysis category, but might also be used within a market

---

[12] Garcia and Bray, 1997.

[13] Danielle Bütschi, Rainer Carius, Michael Decker, Søren Gram, Armin Grunwald, Petr Machleidt, Stef Steyaert, and Rinie van Est, "The Practice of TA: Science, Interaction, and Communication," in Michael Decker, Miltos Ladikas, Susanne Stephan, and Friederike Wütscher, eds., *Bridges Between Science, Society and Policy*, Berlin: Springer, 2004.

[14] Thien A. Tran, "Review of Methods and Tools Applied in Technology Assessment Literature," *PICMET '07— 2007 Portland International Conference on Management of Engineering & Technology*, Portland, Ore., August 5–9, 2007; Thien A. Tran, and Tugrul Daim, "A Taxonomic Review of Methods and Tools Applied in Technology Assessment," *Technological Forecasting and Social Change*, Vol. 75, No. 9, 2008; and Kang, Daekook, Wooseok Jang, Hyeonjeong Lee, and Hyun Joung No, "A Review on Technology Forecasting Methods and Their Application Area," *International Journal of Mechanical, Industrial and Aerospace Sciences*, Vol. 7, No. 4, 2013.

FIGURE 3.1

Decision Tree to Help Identify Type of Technology Assessment Concept to Use

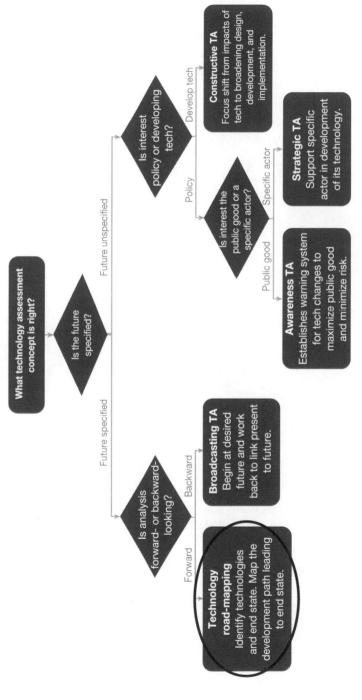

NOTES: Red circle indicates technology assessment concept used in the research described in this report. TA = technology assessment.

analysis (e.g., to consider costs and benefits of market regulations in the coal and gas industry).[15]

Limitations of Scenario-Based Assessments

The research questions associated with this report fall within Figure 3.1's mapping to the technology road-mapping concept. We loosely follow the typical Garcia and Bray approach,[16] with some adaptations for the specific research questions relevant for the Army. The adaptation we employed includes scenarios to help identify technology requirements and capability gaps. Here, we provide a short description of the pros and cons of scenario-based assessments.

Scenario analysis is the process of selecting several scenarios that systematically sample across variables, and then investigating the behavior of those scenarios.[17] Most often, the researcher will hypothesize which parameters are most likely to affect the research question, and the scenarios will describe the likely range of those parameters.[18]

Scenarios are helpful when the full world is too big for any one person to consider at a time. Through worldbuilding, the researcher can help add color to the specifics and provide a common understanding that stakeholders can anchor on to help them consider potential specifics. This allows the research participants to be able to better ingest the research questions, which could, for instance, allow them to better mesh concepts and questions

---

[15] Daniel Gonzales, Timothy R. Gulden, Aaron Strong, and William Hoyle, *Cost–Benefit Analysis of Proposed California Oil and Gas Refinery Regulations*, Santa Monica, Calif.: RAND Corporation, RR-1421-DIR, 2016.

[16] Garcia and Bray, 1997.

[17] William R. Huss, "A Move Toward Scenario Analysis," *International Journal of Forecasting*, Vol. 4, No. 3, 1988.

[18] Pei Hsia, Jayarajan Samuel, Jerry Gao, David Kung, Yasufumi Toyoshima, and Cris Chen, "Formal Approach to Scenario Analysis," *IEEE Software*, Vol. 11, No. 2, 1994; Hannah Kosow and Robert Gaßner, *Methods of Future and Scenario Analysis: Overview, Assessment, and Selection Criteria*, Bonn, Germany: German Development Institute and Deutsches Institut für Entwicklungspolitik, 2008.

with their mental models and thus improve fidelity of response. This also allows the researcher to better investigate uncertainties in the data.[19]

The caveat with scenario building is that, by design, this method means the researcher is only considering a subset of the world, and the process is dependent on the specific scenarios and characteristics under assessment. Consideration must be taken to consider ahead of time what factors are potentially most important to the researcher, and thus ensure that the scenarios used reflect the range of likely outcomes of those (typically exogenous) factors. If it is unclear what characteristics matter most to the research question, then it will be extremely difficult to ensure the scenarios span the dimensions of the research question. Furthermore, in cases where these characteristics are unclear, or probabilities of different characteristics unknown, it may be better to choose a method that considers a larger range of future probabilities (e.g., robust decisionmaking).[20]

Scenario-Based Technology Assessment Approach

We proceeded with a scenario-based technology assessment that builds from existing literature from Garcia and Bray. The specific steps involved in this method are presented in Figure 3.2 and further described next.

The approach begins with the scenarios that we developed through the process discussed in the prior chapter. A key component of these scenarios is a mission overview, which contains the sequential activities necessary to accomplish the mission. For each of these activities, a set of specific challenges must be overcome to accomplish the mission, and Step 3 involves working with subject-matter experts (SMEs) to articulate these challenges in detail.[21]

---

[19] Theo J. B. M. Postma, and Franz Liebl, "How to Improve Scenario Analysis as a Strategic Management Tool?" *Technological Forecasting and Social Change*, Vol. 72, No. 2, 2005.

[20] Robert J. Lempert, "Robust Decision Making (RDM)," in Vincent Marchau, Warren Walker, Pieter Bloemen, and Steven Popper, eds., *Decision Making Under Deep Uncertainty*, Cham, Switzerland: Springer, 2019.

[21] The Step 3 process we used follows the "Strategies-to Tasks Framework" outlined in Robert S. Tripp, Kristin F. Lynch, John G. Drew, and Robert G. DeFeo, *Improving*

FIGURE 3.2
Our Scenario-Based Technology Assessment Approach

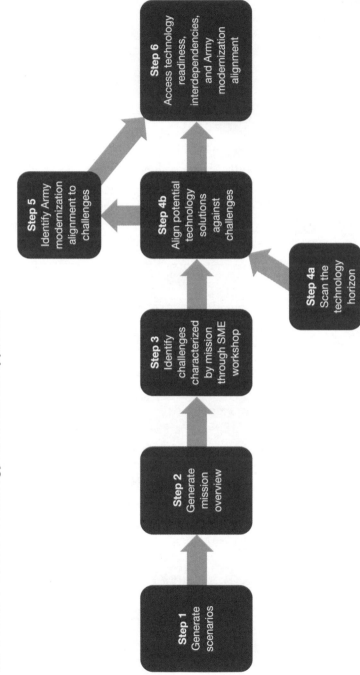

In general, the expertise required to support this step include detailed understanding of the military operations described in the mission overview and knowledge of potential technology areas that may be applied. Both the operational and technology considerations of the scenario may present challenges within the mission space mapped in this step. Step 3 involves its own set of sequential steps that start with identifying the objectives of each mission step (e.g., transportation to theater, base defense, billeting, attack). Once this objective is identified at a high level, we outlined a set of subobjectives that may support the primary mission step objective (e.g., land and sea transportation, different types of base defense). These objectives are then aligned with warfighting functions (WFF) that represent an action area for the Army (e.g., movement and maneuver, intelligence). We also identified tasks specified by the mission overview step that aligns with objectives and actions (e.g., move 11,600 soldiers approximately 1,200 miles). Lastly, for these combinations of objectives, actions, and tasks, we identified the key challenges associated with the different alternatives available.

Alongside this work with the scenario, we conducted an independent scan of the technology horizon. This scan produced a set of technology research areas across multiple domains (such as the physical domain, biological domain, and information domain). These technology research areas include

- directed energy
- hypersonics
- energy
- food and water security

Air Force Command and Control Through Enhanced Agile Combat Support Planning, Execution, Monitoring, and Control Processes, Santa Monica, Calif.: RAND Corporation, MG-1070-AF, 2012. In implementing the Step 3 process for the Battle of the Arctic Depths scenario, SMEs were identified from our study team. These SMEs had expertise in technology forecasting, military logistics, operations, and specific technology areas (such as space, biotechnology, artificial intelligence [AI], and energy). The expertise required to identify challenges in Step 3 included SMEs with a background in both the military operations described in the scenario and the potential solution space for conducting the scenario. A total of eight researchers were present at the SME workshops, which were held over approximately four hours for the Battle of the Arctic Depths scenario.

- materials and manufacturing
- mass transportation
- quantum technologies
- autonomous systems and robotics
- computer hardware and chips
- position, navigation, and timing (PNT)
- space technologies
- biotechnology
- synthetic biology
- AI
- big data analytics
- communication technologies
- digital security
- information and perception manipulation
- virtual and augmented reality.

In Step 4b, we worked to align these technologies as potential solutions to the challenges identified in Step 3.

In Step 5, we identified current Army modernization efforts to also align those efforts to the challenges.

Finally, in Step 6, we assessed the technology according to several dimensions. These include the overall readiness of the technology interdependence between the technology development, and the extent that Army modernization priorities align with the identified challenges.

Technology Assessment for the Battle of the Arctic Depths Scenario

To demonstrate how this technology road-mapping process may be applied to the developed scenarios, we will overview the process against the Battle of the Arctic Depths scenario. This discussion will first introduce the developed scenario, review the mission overview, demonstrate how challenges have been identified, align technologies and Army modernization priorities to those challenges, and finally provide an assessment of candidate technology solutions. For a more detailed discussion of all developed scenarios, see Appendix A.

Introduction to the Battle of the Arctic Depths Scenario

In the Battle of the Arctic Depths, climate change has melted the Arctic region's protective shield of ice and opened its vast lands, seas, and plethora of natural resources to human activity. States with territorial claims in the Arctic—and key near-Arctic nations—do not agree on how these resources should be divided, with many proffering overlapping claims. China maintains that the region is one of undetermined sovereignty. Denmark claims an area extending from Greenland past the North Pole to the limits of the Russian Exclusive Economic Zone.[22] The conflict is initiated under the guise of a scientific exploration trip during which Chinese forces plant a flag and claim sovereignty over a 300,000 km^2 swath of Arctic territory, all of which is claimed by Denmark. As a result of China's attempts to bolster sovereignty through "facts on the ground," the North Atlantic Treaty Organization (NATO) invokes Article 5 to forcibly remove the People's Liberation Army (PLA) from the contested area. Thus, the United States and its NATO allies find themselves in armed conflict with a near-peer state in the Arctic.

This scenario presents key challenges for the Army. First, the Army must establish an expeditionary presence without a sustained logistics tail in an austere, unpredictable environment, which poses a challenge to supplying fuel and power and equipment maintenance. Additionally, the environmental conditions in the Arctic present a uniquely biologically hazardous environment. To respond to this crisis, the Army—along with Joint Interagency, Intergovernmental, and Multinational (JIIM) partners—must operate as a combined force employed in an expeditionary manner, maneuvering long distances in an area characterized by extreme cold weather and high-altitude conditions to prevent a near-peer adversary from solidifying territorial claims.

---

[22] United Nations, Division for Ocean Affairs and the Law of the Sea, "Commission on the Limits of the Continental Shelf (CLCS), Outer Limits of the Continental Shelf Beyond 200 Nautical Miles from the Baselines: Submissions to the Commission: Submission by the Kingdom of Denmark," November 2, 2015.

Battle of the Arctic Depths Mission Overview

Before identifying the challenges associated with the scenario, as a part of the technology road-mapping process, we have specified the operational mission. A mission overview allows the scenario to be expressed in terms of identifiable mission steps that will lead to mission success based on the problem statement specified in the scenario. For the Battle of the Arctic Depths scenario, the mission entails a need to deploy forces into the Arctic region without a sustained logistics tail, establish basing operations, develop a common operational picture (COP) to plan offensive operations, expel Chinese forces from the region, and, ultimately, redeploy forces once all mission objectives have been accomplished. These steps form the basis of the scenario decomposition that allows challenges to be identified and expressed concretely as follows:

1. After 30 days of uncontested Red actions in the region, Blue U.S. forces receive a notice to proceed with movement of assigned forces in Alaska to establish a tactical assembly area (TAA) within operational proximity of the contested territory.
    a. This step includes two brigade combat teams and two aviation battalions of roughly 11,600 soldiers at Fort Wainwright, Joint Base Elmendorf-Richardson, and Fort Greely. The distance to move these forces is approximately 1,200 miles.
2. Upon arrival at the TAA location, U.S. forces will establish basing operations in an extremely austere environment. Frequent extreme weather, difficult terrain, and the slow spread of bubonic plague throughout the area pose significant challenges to operations. Despite these factors, U.S. forces are responsible for rapidly establishing the following for the JIIM:
    a. base defense
    b. billeting
    c. food and water
    d. prime power.
3. Within 20 days of the notice to proceed, Allied forces will begin arrival and integration at the TAA. This will include additional forces from the United Kingdom, Germany, and Denmark of an

approximately equivalent size as U.S. forces establish operations. U.S. forces are responsible for supporting

a. secure communications

b. reception, staging, onward movement, and integration of NATO forces

c. transition of command and control to NATO commanders

4. Blue forces will establish a COP of Red through unified intelligence, surveillance, and reconnaissance (ISR) operations. Red may employ advanced techniques to mask radio frequency signatures of equipment and take advantage of the extreme weather and mountainous terrain to counter Blue intelligence-gathering. Throughout ISR operations, Blue will operate without a sustained logistics tail, responsible for maintaining TAA operations with infrequent and unpredictable resupply. Successfully establishing a COP will require Blue forces to

a. determine the location and size of Red forces in the region (i.e., a full division of Red forces may have been deployed but must be confirmed, and the posture determined)

b. identify offensive and defensive capabilities of Red

c. establish a pattern of life for Red activities, including daily operations, movements of forces, resupply missions, and defensive capabilities.

5. Blue forces will coordinate and execute an offensive mission against Red to expel them from the region. This effort will continue to be accomplished with an infrequent and unpredictable logistics tail. Additionally, continued ISR of Red will be required to provide a COP to mission commanders.

6. Upon expulsion of Red forces, Blue forces will establish a continued presence in the previously occupied area. Blue will continue ISR operations monitoring the entire area occupied by Red forces. Ground truth will need to be established over an extended period of time to ensure successful expulsion of Red.

7. Host-nation forces allied with Blue will redeploy the majority of forces to their home stations. A small presence of host-nation forces may be kept within the previously occupied area, and remote ISR

will continue monitoring to ensure no further Red activities in the region.

Identification of Challenges in the Arctic

With a mission overview specified, each of the identified mission steps can be analyzed to establish the operational challenges at play. To identify these challenges, we have used a process reflective of the Strategies-to-Tasks framework.[23] This framework provides a structured process to identify objectives of the mission overview step, decompose those objectives into subobjectives, align them with WFF that represent actions for the Army, identify the tasks required to complete these actions, and, ultimately, associate these tasks with the combinations of objectives, actions, and tasks. This framework can be executed through the following series of steps:

1. Identify the objectives of the mission step (e.g., transportation, base defense, billeting).
2. Outline subobjectives that may support primary mission objective (e.g., land and sea transportation).
3. Align objectives with the WFF representing an action area for the Army (e.g., movement and maneuver).
4. Identify tasks specified by mission overview step that align with objectives and actions (e.g., move 11,600 soldiers approximately 1,200 miles).
5. For these combinations of objectives, actions, and tasks, identify the key challenges associated with each alternative (e.g., air vehicles must be able to land without established runways).

Implementing this Strategies-to-Tasks framework for the mission overview steps identified previously provides the structure to directly tie challenges to the Battle of the Arctic Depths scenario. These challenges serve as the input to subsequent steps in the technology assessment process where candidate technology solutions and Army modernization priority alignment are assessed to address specific challenges. In implementing the Step 3

---

[23] Tripp et al., 2012, pp. 99–110.

process for the Battle of the Arctic Depths scenario, SMEs were identified from the study team. These SMEs had expertise in technology forecasting, military logistics, operations, and specific technology areas (such as space, biotechnology, AI, and energy). The expertise required to identify challenges in Step 3 included SMEs with a background in both the military operations described in the scenario and the potential solution space for conducting the scenario. A total of eight researchers were present at the SME workshops, which were held over approximately four hours for the Battle of the Arctic Depths scenario. A snapshot of the resulting analysis for this Arctic Scenario is provided in Table 3.1. As is demonstrated in this table, mission steps, such as deployment to a TAA, can be decomposed into the objective of transportation by land and sea, requiring the action of movement and maneuver, specifically for a large number of soldiers and their equipment. This presents challenges of providing vehicles with great enough efficiency and lift capability to strategically move these forces. These challenges will then ultimately support the identification of potential technology solutions.

Technology Alignment to Arctic Depths Challenges

A key outcome of the technology road-mapping process is the identification of candidate technology solutions intended to address scenario-based challenges. In this step of the process, the analyst must consider the challenges to identify the pertinent technology domains and research areas where technology solutions may lie. This activity is directly supported by the technology horizon scan, discussed earlier. Further analysis within these technology domains and critical research areas is used to identify the candidate technology solutions if the portfolio of technologies currently being tracked does not contain an adequate technology solution. Operationalizing this step in the technology road-mapping process can be accomplished through the following steps:

1. Considering the challenge, identify the pertinent technology domain(s) and research areas where technology solutions may exist
2. Identify candidate technologies from information on technologies collected in those domain(s) and research areas

TABLE 3.1

Snapshot of Battle of the Arctic Depths Strategies-to-Tasks Challenge Identification

Mission Step	Objective	Subobjective	WFF (Action)	Task	Challenge
Deploy to TAA	Transportation	Land and sea	Movement and maneuver	Move 11,600 soldiers, equipment, and sustainment supplies about 1,200 miles	Land: Engine efficiency drops as vehicles experience engine power reductions by 4–6 percent for every 1,000-meter increase in elevation above sea level (mileage and load carrying capacity are reduced by 25 percent and need for fuel and oil increases 30–40 percent)
					Land: Vehicles incur higher rates of mechanical problems because of thickened oil and lubricants because of low temperatures
		Air and space			Air: High altitudes restrict helicopter lift capabilities and decrease aircraft payloads
					Air: Vehicles need to be able to land without established runways
					Space: Vehicles need to be able to land without established infrastructure

Table 3.1—Continued

Mission Step	Objective	Subobjective	WFF (Action)	Task	Challenge
Establish a COP	ISR	Physical data collection	Intelligence	Collect, analyze, and assess information in real time; share information across partners	Masking effect (mountainous terrain blocks radar beams)
					Radar signal scattering may occur because of ice, fog, and airborne snow
		Real-time decisionmaking			Big data analysis is time- and resource-intensive
Establish Blue presence	Occupying presence	Continued presence with soldiers	Protection	Make medical facilities available to support 11,600 soldiers, responding to threats from the enemy and environment	Potential increase in infectious diseases, whether from limited water and confined living spaces or from diseases being released from melting permafrost
					Potential increase in infectious diseases because personal hygiene difficult to maintain because of limited water and confined living spaces

3. If none are present in existing market research, conduct novel research and/or solicit information from SMEs on candidate technologies in the identified domain(s) and research areas

Implementing these steps with the previously identified process allows candidate technology solutions to be identified for the Battle of the Arctic Depths scenario. These candidate technology solutions, aligned with the associated challenges, are presented in Table 3.2. As can be observed, challenges such as the need for efficient engine use may require advanced lubricant technologies, while others, such as the need for high-capacity air lift, may require such technologies as vertical take-off and landing (VTOL) air assets. For this particular scenario, the key technologies identified for these challenge areas included lubricants, quantum, space transport, air vehicles, autonomous weapon systems (AWS), and biotechnology.

TABLE 3.2
Battle of the Arctic Depths Technology Alignment

Mission Step	Challenge	Technology Alignment
Deploy to TAA	Land: Engine efficiency drops as vehicles experience engine power reductions of 4–6 percent for every 1,000-meter increase in elevation	Use proper motor oil to improve efficiency; electric vehicles (which have greatly improved engine efficiency compared with gas-powered)
	Land: Vehicles incur higher rates of mechanical problems because of thickened oil and lubricants from low temperatures	Use proper motor oil and lubricant to avoid increased viscosity; electric vehicles
	Air: High altitudes restrict helicopter lift capabilities and decrease aircraft payloads	Future vertical lift (FVL); drones/unmanned aerial vehicles (UAVs)
	Air: Vehicles need to be able to land without established runways	VTOL
	Space: Vehicles need to be able to land without established infrastructure	Earth point-to-point (P2P)

Table 3.2—Continued

Mission Step	Challenge	Technology Alignment
Establish a COP	Masking effect (e.g., mountainous terrain blocks radar beams)	Radar signal processing; quantum technologies
	Radar signal scattering may occur because of ice, fog, and airborne snow	Radar signal processing; quantum technologies
	Big data analysis is time- and resource-intensive	Advanced computing, quantum computing, AI, and machine learning; virtual/augmented reality for enhanced data visualization
Establish Blue presence	Potential increase in infectious diseases, due to diseases being released from melting permafrost	Biomaterials, air-to-water technologies, portable water purification, water reuse systems
	Potential increase in infectious diseases because personal hygiene is difficult to maintain	Biomaterials, air-to-water technologies, portable water purification, water reuse systems

Army Modernization Priority Alignment in the Arctic

In addition to understanding the potential candidate technology solutions, it is also important to consider what actions the Army, or any of its partners, are taking to address the challenges and their associated technology solutions. In this way, Army modernizations priorities are evaluated according to the challenge areas and the candidate technology solutions to address any potential gaps or misalignment. Accomplishing this requires a broad look at Army R&D, acquisition, and other activities aligned with the technology areas. This may be accomplished by considering the challenges and candidate technology solutions while identifying ongoing efforts throughout the Army and its partners. The following steps support are used:

1. Considering the challenge and candidate technology solutions, identify Army modernization priorities and ongoing research efforts, including
    a. examining priorities in CFTs and across Army priority research areas

    b.   identifying ongoing Army R&D efforts within technology domains and research areas.

2.   Provide an accounting of ongoing efforts aligned to challenges and candidate technologies.

This alignment of Army modernization priorities and activities has been generated for the challenges and candidate technology solutions identified previously (see Table 3.3). As an example of how this alignment occurs, consider the challenge associated with the higher rate of mechanical problems from thickened oil at low temperatures. The candidate technology solution in this case included advanced lubricants designed specifically for cold temperatures. Within the Army, U.S. Army Tank Automotive Research, Development, and Engineering Center (TARDEC) is currently developing a single common powertrain lubricant (SCPL). This activity represents an alignment of an Army modernization priority to this challenge and technology area. However, it is important to note that the intent at this step is to provide an accounting of all Army modernization priorities aligned to challenges and candidate technology solutions. It is not until the next step in the technology road-mapping process that an assessment is made of the benefit such alignment may or may not have.

Assessment

The assessment of technology alignment and readiness is based on the following four questions.

1.   Is the technology projected to have a technology readiness and manufacturing readiness sufficient for implementation in the time frame considered? Alternatively, what is the expected technology/ manufacturing readiness in the time frame considered?
2.   What interdependencies exist for the technology?
3.   Where is development of the technology expected to occur? (i.e., will this be a commercial technology, dual use, or military use?)
4.   How are Army modernization priorities aligned to the technology and any existing interdependencies?

TABLE 3.3

Battle of the Arctic Depths Army Modernization Alignment

Mission Step	Challenge	Army Modernization Alignment
Deploy to TAA	Land: Engine efficiency drops as vehicles experience engine power reductions by 4–6 percent for every 1,000-meter increase in elevation	Advanced Combat Engine in development at TARDEC; Advanced Powertrain Demonstrator initiative; Joint Operational Energy Initiative; Tactical Vehicle Electrification Kit
	Land: Vehicles incur higher rates of mechanical problems because of thickened oil and lubricants from low temperatures	SCPL at TARDEC (testing included the Arctic); fuel-efficient gear oil (FEGO); science of additive manufacturing as a priority research area
	Air: High altitudes restrict helicopter lift capabilities and decrease aircraft payloads	FVL (modernization priority)
	Air: Vehicles need to be able to land without established runways	FVL (modernization priority); many (primarily small) UAVs have VTOL capability, and more are in development; autonomy research as a priority area
	Space: Vehicles need to be able to land without established infrastructure	No Army-specific capability; U.S. Space Force/U.S. Air Force Research Laboratory (AFRL) Rocket Cargo experimental program

Table 3.3—Continued

Mission Step	Challenge	Army Modernization Alignment
Establish a COP	Masking effect (mountainous terrain blocks radar beams)	Quantum as a priority research area
	Radar signal scattering may occur because of ice, fog, and airborne snow	Assured PNT modernization priority; quantum as a priority research area
	Big data analysis is time- and resource-intensive	Quantum, AI, and automation are priority research areas
Establish Blue presence	Potential increase in infectious diseases because of diseases being released from melting permafrost	Synthetic biology as a priority research area
	Potential increase in infectious diseases because personal hygiene is difficult to maintain	Synthetic biology as a priority research area

For the Battle of the Arctic Depths scenario these questions are addressed for technology areas of lubricants, quantum, space transport, air vehicles, AWS, and biotechnology. These technology areas are derived directly from the technology and Army modernization priority alignment steps previously discussed. A summary of this research is presented in Table 3.4. In many instances the Army is well aligned to the technology areas, especially with respect to the advances in quantum, AWS, and biotechnology. However, misalignment is observed in some technology areas, such as lubricants alluded to earlier, that may need to be addressed through investment of resources, changes in priority, and the like. The one area where no Army priority alignment was observed was with respect to space transport. This is likely due to the fact that the development of the technology is occurring exclusively in the commercial sector, and any integration into a DoD service would likely occur with one of the Army's sister services.

Although Table 3.4 provides a general overview of the assessment of technologies and Army modernization priority alignment to the identified challenges, we offer brief additional discussion across each of these technologies. For more information on these technology areas, see Appendix C.

Lubricants

Lubricants reduce friction, wear, and energy consumption and are subsequently required to operate every piece of machinery. In cold weather, viscosity becomes of great concern, specifically as it relates to lubricants. As a result, in recent years, R&D related to lubricants, particularly in extremely cold conditions (e.g., the Arctic), has increased, resulting in a host of lubricants and greases with ideal properties. Both the automotive and maritime industries are investing in this space—particularly the latter, given the presence of Arctic shipping routes, because such lubricants exhibit a high state of readiness with development occurring in the commercial sector for dual use in military applications.

In addition to increasing interest from several public sectors, the Army has invested in the development of new lubricants.[24] Testing of a SCPL—a synthetic, all-season, fuel-efficient, heavy-duty engine oil developed at the

---

[24] The investments are through the Ground Vehicle Systems Center (GVSC)'s Fuels and Lubricants Research Facility (located at Southwest Research Institute in Texas) and the

TABLE 3.4

Battle of the Arctic Depths Technology and Army Modernization Assessment

Technology	TRL/MRL[a] (Low/ Medium/High)	Interdependency (None/ Few/Many)	Development Area (Commercial/ Dual/MIL)	Army Alignment (No Alignment/ Mismatches/ Strong)	Key Issue
Lubricants	High	None	Dual	Mismatches	Army single spec lubricant ignores specialized lubricant development from commercial sector.
Quantum	Low/Medium	Many (possible)	Dual	Strong alignment	Significant uncertainty and disagreement among experts on feasibility of quantum technologies within specific time frames.
Space transport	Low	Many	Commercial	No alignment	Readiness of technology is questionable. Only a single commercial company currently working on development with limited commercial business case makes P2P space travel risky without significant DoD investment.
Air vehicles	High	Many	MIL/Commercial	Mismatches	Army development occurring on light and medium payload VTOL, and it is lacking for heavy VTOL.

Table 3.4—Continued

Technology	TRL/MRL[a] (Low/ Medium/High)	Interdependency (None/ Few/Many)	Development Area (Commercial/ Dual/MIL)	Army Alignment (No Alignment/ Mismatches/ Strong)	Key Issue
AWS	High	Many	MIL	Strong alignment	Ethical, legal, and policy constraints and risks of reliability and mistrust may limit the use of lethal autonomous weapon systems (LAWS).
Biotechnology	High	Many	All	Strong alignment	A need to further advance technology and develop new technology, but many pertinent technologies are mature enough for ready deployment.

NOTES: MIL = military.

[a] Technology Readiness Level (TRL) and Manufacturing Readiness Level (MRL) are measures of the operational and manufacturing maturity of the technology, respectively. These metrics are assessed on a scale from 1 to 9, with 1 representing a technology that is still preconceptual in design and 9 representing a commercial off-the-shelf technology. Because of the difficulty in providing a direct assessment of the TRL and MRL for technology areas, a scale with low, medium, or high is applied here. A low TRL and MRL are associated with levels typically in the 1-to-3 range, which implies low technical and manufacturing maturity, typically still in design. A medium rating would correspond with TRL and MRL from 4 to 6 generally, in which operational prototypes and bespoke manufacturing may be present. A high rating suggests relatively high maturity of the technology, with TRL and MRL ratings of 7 to 9, in which operational effectiveness and manufacturing readiness has been observed or is likely to be feasible in the near term.

GVSC research facility—was focused specifically on the Arctic. Therefore, it is likely that the required specifications under the SCPL will meet the challenges outlined for the Arctic, although a single specification lubricant may operate less efficiently in other operating environments. Therefore, it may be of interest to further study the logistics burden of housing multiple lubricants designed for specific applications to take advantage of the advances in this technology.

Quantum Technologies

Quantum science combines elements of mathematics, computer science, engineering, and physical science to study photons and electrons—the smallest particles of matter and energy. Quantum technologies can be broken down into three applications—computing, communications, and sensing. Quantum computing will pose a distinct threat to systems that rely on public key cryptography, rendering all present and future data vulnerable. Quantum communications would provide the ability to conduct communications that cannot be broken, improving the security of communications against interception and eavesdropping.[25] Quantum sensors could also contribute to improved inertial navigation systems, allowing submarines to operate without active sonar (thereby complicating the adversary's ability to find and attack submarines).[26]

Not unlike many others, the Army is investing heavily in quantum technologies, both by funding research efforts and establishing new agencies, positions, and programs dedicated to this space. Some examples include the work spearheaded by U.S. Army Combat Capability Development (DEVCOM) and the Army Research Laboratory's Center for Distributed Quantum Information, focused on improving sensing and methods to build

---

Armament Research, Development, and Engineering Center at Picatinny Arsenal in New Jersey.

[25] Patrick Tucker, "China has a Breakthrough in Spy-Proof Quantum Communications," *Defense One*, November 9, 2017.

[26] Alan Cameron, "Quantum Magnetometer Senses Its Place," *GPS World*, May 8, 2019; David Kramer, "DARPA Looks Beyond GPS for Positioning, Navigating, and Timing," *Physics Today*, Vol. 67, No. 19, 2014.

quantum communication networks;[27] and the research hub recently stood up by the Army Research Office and National Security Agency's Laboratory for Physical Sciences, which is part of the National Quantum Initiative—a coordinated federal plan to accelerate and advance quantum technology–centered studies and applications.[28]

It is worth noting that, although garnering significant interest from virtually all sectors and possessing the ability to be truly transformative in the battlespace, even experts wildly disagree on the feasibility of these technologies within certain time frames. In short, quantum advances may perpetually be on the horizon. In the next 20 years, few (if any) of these technologies are likely to advance to the point where they are capable of being deployed in support of combat operations.

Biotechnology

Many different potential biotechnology applications are relevant in this Arctic scenario. In short, biotechnology has the ability to enhance warfighting materiel and systems, optimizing warfighter health and performance and military medicine. The potential use cases for biotechnology in any scenario are countless given the exceedingly broad range of technologies and applications that fall under this heading.

The Army is investing heavily in biotechnology, as are the sister services. Synthetic biology, in particular, remains a priority research area within the Army and across broader DoD. The Army Research Lab has the Transformational Synthetic Biology for Military Environments Essential Research Program, focused on self-assembling and self-healing materials, active camouflage, and on-demand production of small quantity materials. The

---

[27] Todd South, "Quantum Breakthroughs Help Army, Air Force Advance Super-computing," *Army Times*, April 6, 2021; U.S. Army Combat Capabilities Development Command, Army Research Laboratory Public Affairs, "Army-Funded Research Lays Groundwork for Future Quantum Networks," March 11, 2021; Youpeng Zhong, Hung-Shen Chang, Audrey Bienfait, Etienne Dumur, Ming-Han Chou, Christopher R. Connor, Joel Grebel, Rhys G. Povey, Haoxiong Yan, David I. Schuster, and Andrew N. Cleland, "Deterministic Multi-Qubit Entanglement in a Quantum Network," *Nature*, Vol. 590, 2021.

[28] Brandi Vincent, "NSA, Army Launch 'Qubit Collaboratory' to Advance Quantum Information Science," NextGov.com, May 4, 2021.

Army's Engineer Research and Development Center (ERDC) has a history of genetically modifying microorganisms for biological detection and contaminant (e.g., oil spill, munition) degradation. Outside synthetic biology, there is also significant development on external devices, such as exoskeletons, that can augment individual soldier strength and capability in austere environments.

Many of these biotechnologies are presently at a maturation level that allow for deployment in the field in their current state, though further R&D will only continue to improve these technologies while also creating new ones. We recommend that the Army continue investing in biotechnology while simultaneously looking to industry and academia for further advances in the space.

Space Transport for Earth P2P

The space sector is developing space P2P technology for transportation from a point to another point on Earth. Achieving operational capability for space P2P depends on significant developments in related technologies. The commercial sector is working on improving technologies to enhance propulsion and landing technologies; to reduce costs of launch, recovery, and refurbishment; and, perhaps most importantly, to meet safety levels for commercial viability. The accessibility of the technology will heavily depend on the success of commercial P2P business. Without a commercial market, DoD would have to pay substantial price to use space P2P as a mode of transportation.

From the military, the U.S. Space Force and the AFRL are leading the Rocket Cargo Vanguard program to leverage the commercial market's development of space P2P capabilities. Although the U.S. Army Reserves currently have no plans or capabilities for this technology and will rely on traditional logistics/resupply capabilities over land, air, and water for the foreseeable future, the Rocket Cargo Vanguard program can support U.S. Transportation Command's logistics capabilities.

Air Vehicles–Future VTOL

VTOL is a not a new technology. Legacy helicopters and rotorcrafts, such as Chinooks and Black Hawks, have served as VTOL air vehicles for the military for several decades. Key related technologies—such as rotors; propulsion; situational awareness with AI and advanced sensors; avionics, communication, and advanced composite materials—are currently making advances.

Army modernization efforts categorize VTOL in five capability sets, from light to heavy, through the FVL program.[29] Among the capabilities, Army's current focus is on light- and medium-transport VTOL platforms. Capability sets 4 (heavy lift) and 5 (ultra-heavy lift) are more relevant to the transportation challenges presented in the Arctic scenario considering the significant number of troops, equipment, and supplies involved. The time frame for the initial operational capability for the two capability sets are uncertain, with unofficial estimates going out to 2060s.

Although VTOL technology is certainly attractive for a mission similar to the one presented in the Arctic scenario, development is lacking on the heavy and ultra-heavy payload vehicles that would be relevant for the transportation challenges presented in the Arctic scenario.

AWS

Current policies separate offensive AWS, also called *lethal autonomous weapon systems* (LAWS), from defensive AWS. The international community does not have consensus on the development LAWS, mainly due to the ethical considerations around such weapon systems. Despite the lack of international consensus, developments in the required technology are continuing. Many technologies for LAWS derive from technologies in nonlethal autonomous systems, which are not subject to the same ethical criticism at home and abroad. Future development of AWS will be based on evolutionary improvements of the current technologies.

The Army is actively pursuing technologies related to AWS. Investment in technology development includes research in semiautonomous and autonomous vehicles and weapons systems, though human judgment for targeting and firing capabilities is still the main operating concept.

---

[29] Jeremiah Gertler, "Army Future Vertical Lift (FVL) Program," Congressional Research Service, IF11367, last updated July 13, 2021.

CHAPTER FOUR

Insights and Recommendations

As demonstrated in this report, a scenario-based technology assessment can be structured to provide a repeatable process to effectively decompose scenarios into associated challenges, identify potential candidate technology solutions to those challenges, and, ultimately, assess the alignment of Army modernization priorities against those candidate technology solutions and challenges. In conducting this research, insights have emerged both on the scenario for the Arctic Depths discussed here and on the broader development of a technology assessment process to support such an activity. This chapter will present insights and recommendations in both areas, beginning with insights and recommendations associated with the Arctic Depths scenario and, second, presenting the insights and recommendations on conducting technology assessments more generally.

Before reviewing the insights generated from the Arctic Depths scenario, we will note one overarching finding related to scenario generation gleaned over the course of this work. The scenarios that were developed (for an overview, see Appendix A) occurred in contexts in which the United States did not possess a technological advantage, as technology was revolutionary throughout the world, and the United States faced extremely difficult operational challenges. **Even in a technologically advanced future, the Army will face many existing and long-standing difficulties with respect to logistics support of forces, transportation and distribution, ambiguous competition and conflict objections, and significant information competition.**

Arctic Depths Technology Assessment Insights and Recommendations

Key Findings from the Arctic Depths

In exploring the Arctic Depths scenario, **we found that many Army modernization priorities are aligned to the key challenges that the Army would face.** In particular, we assessed that three Army modernization priorities aligned with the technology requirements in the scenario: AWS, biotechnology, and quantum technologies. The alignment in these specific cases reflects certain specific attributes of these three technology areas—the technology application is driven by a military need, it occurs in areas where specific applications can be rapidly developed, or when the technology replicates and improves an existing capability.

However, some current Army modernization priorities exhibited misalignment or no alignment for the identified challenges in the Arctic Depths scenario. The two specific cases where misalignment was observed included technology areas around lubricants and VTOL air vehicles. For lubricants, the Army is pursuing a single specification lubricant while development in the commercial sector is advancing lubricant technologies for specific operating environments. In these cases, the Army misalignment may result in an inability to take advantage of commercial advancements that exist. The misalignment around VTOL was specific to the Arctic Depths scenario, which would require heavy lift capabilities. Given that the Army is currently focused on VTOL advances in the light- and medium-duty lift air assets, there is a potential capability gap. In this case, immediate needs are likely driving current priorities, so this misalignment may be an artifact of the specific scenario explored. However, this misalignment should still be considered because not addressing future heavy vertical lift may introduce risk for expeditionary scenarios in austere environments.

The assessment of the Arctic Depths scenario found that there was no alignment between Army priorities and the potential need for P2P space transportation. This factor does not necessarily mean that something is lacking in Army modernization; for these cases, development may be driven by commercial interests or by other services with a different mission set. For those technologies where commercial development leads innovation, the resulting R&D efforts, applications, or products may not best serve the

Army's needs generally. Commercial products may be designed for specific uses or populations that may not match the military's uses or populations. If this situation appears to be the case, the Army may need to either develop its own options, ensure other services are developing the capability, or work with commercial industry to produce a more appropriate solution. Before engaging with private industry in the development of P2P space transport, the Army should formalize and develop requirements for the capability based on expected military needs in the future. Further analysis of the expected mission space will be required to determine the criticality of such a technology for mission success.

Finally, **we found that there are unlikely to be purely technology-based solutions to many challenges, and technology itself potentially introduces new risks.** Investment in technologies may have ancillary benefits and costs that should also be considered. As an example, broad employment of AWS is likely to significantly affect operational energy needs, which would have implications for the delivery and availability of onsite energy generation. Additionally, the proliferation of AWS may also result in new risks as adversaries may equally leverage these technologies in the context of future crises. These downstream effects of technologies may require further assessment beyond the scope considered here.

Recommendations from the Arctic Depths

In conducting the technology assessment for the Arctic Depths scenario, we identified specific cases of misalignment of Army modernization priorities and technology development. In these instances, the Army may consider the following recommendations:

- The Army should develop key partnerships in the private sector and with multinational partners to ensure the availability and integration of technology it needs, especially for the technologies identified as critical needs through technology assessment.
- Where misalignment in technology development and Army modernization priorities exist, the Army should consider investing resources, updating doctrine, and creating alternative strategies to take advantage of the required technologies.

In cases in which the emerging needs from technology assessment are not aligned to Army modernization priorities, such as space transport in the Arctic Depths scenario, it is likely that the development will be dependent on commercial partners for the capability. When Army modernization priorities are misaligned to technology development needs because another service or military component leads that development, the Army should prepare joint relationships to ensure that it has access and can use that advancement when it is mature. Army modernization benefits from active and comprehensive coordination of research, development, and deployment across industry partners, sister services, and components. Additionally, if one or more of these orthogonal technologies are of interest for the Army, additional effort will have to be made for the Army to support the specifications or operational requirements that will make them an effective tool.

In instances where there is misalignment of current Army modernization priorities and emerging needs identified through technology assessment, such as those for lubricants and VTOL, the Army may consider investment of resources to realign priorities if employment of the technology is deemed critical to mission success. This investment may require (1) changes to planned acquisition strategies, such as moving from single specification lubricants to logistics strategies aligned with specialized lubricant acquisition for specific operating environments, or (2) a shift in focus, such as for VTOL to heavy capacity air assets. These changes are likely to have downstream effects that will need to be studied to fully account for the required investments.

Finally, it should be noted that the scenario development activities pursued under the current scope of work were ultimately limited to future world states where revolutionary technology advancements were considered. To provide a more balanced perspective across scenario-based technology assessments, the Army should further develop this portfolio of scenarios to include scenarios in which technological advancements are evolutionary in nature. This would allow the Army to provide a broader perspective from technology assessment across the exogenous uncertainty of technology development.

Additional Considerations for the Army

Transformational technological changes, such as those anticipated by our analyses, will likely cause a need for changes to organizational structures, doctrine, personnel, interoperability needs, or WFF. Past transformative technologies, such as cyber, unmanned aerial systems (UASs), and AI, have been disruptive to Army operations at varying levels and may be useful for identifying generalizable lessons for other future technologies. For instance, cyber capabilities have been extremely transformative, requiring an entirely new command for accommodation. However, UAS have substantially expanded Army warfighting capabilities in intelligence but have required few changes to organizational structure or WFF. In this way, the Army might need to consider how candidate technology solutions in the future affect interoperability; WFF; and doctrine, organization, training, materiel, leadership and education, personnel, and facilities.

Cyber, AI, and UAS have presented significant challenges to interoperability, as many other computing-based technologies will in the future. Specifically, UASs have had challenges with joint interoperability, and no joint doctrine on the management and employment of UASs remains.[1] Much of the challenge for these technologies, and other future technologies in which interoperability with partners will exist, lies in the inherent trade-off between interoperability and security, especially with data-sharing. Doctrine, policy, and processes govern this trade-off and need to be developed with future operational success in mind.

Future technologies may affect WFF, but past implementations of cyber, AI, and UASs have not necessitated significant changes. Even for the most-disruptive cyber technologies, the existing six WFF are expected to be used as designed in traditional operations for integration, synchronization, and

---

[1] Bernard D. Rostker, Charles Nemfakos, Henry A. Leonard, Elliot Axelband, Abby Doll, Kimberly N. Hale, Brian McInnis, Richard Mesic, Daniel Tremblay, Roland J. Yardley, and Stephanie Young, *Building Toward an Unmanned Aircraft System Training Strategy*, Santa Monica, Calif.: RAND Corporation, RR-440-OSD, 2014; Curtis L. Blais, *Unmanned Systems Interoperability Standards*, Monterey, Calif.: Naval Postgraduate School, NPS-MV-16-001, September 2016.

command and control.[2] Each WFF will be affected to a different degree by the integration of such technologies as AI on the basis of the need for and utilization of detection, rapid decisionmaking, and automation.[3] Generally, investment in a technology should roughly follow the potential advancement in the WFF as a result of integration.

Technology Assessment Process Insights and Recommendations

The technology assessment for the Arctic Depths enabled us to uncover insights and recommendations regarding specific technologies; the development of the technology assessment process itself should also be viewed as a contribution of this research. This process may provide a structured approach for future Army activities, such as tabletop exercises, workshops, and working groups.

Key Findings on Technology Assessment

As discussed in Chapter Three of this report, many technology assessment concepts could be used for assessment activities based on the type of information available and desired outcome of the activity. Selection of the appropriate concept to employ should be based on the purpose of the information generated during a technology assessment. For instance, if technology assessments are intended to understand policy impacts of a specific actor without specifying the future a priori then a strategic technology assessment concept would be the most appropriate. For the work conducted here, the desired end state of the technology assessment was to provide a forward-looking analysis on how technology may be applied to future scenarios, which necessitated the use of technology road-mapping. As future activities

---

[2] Walter S. Sutton, *Cyber Operations and the Warfighting Functions*, Carlisle Barracks, Pa.: U.S. Army War College, March 2013.

[3] Thomas Ryan, and Vikram Mittal, "Potential for Army Integration of Autonomous Systems by Warfighting Function," *Military Review*, September–October 2019.

are pursued, the Army may want to consider the justification for the assessment to ensure that the appropriate concept is applied.

Pursuing technology road-mapping under this effort required the development of scenarios to provide a specific future against which technologies could be evaluated. The scenarios that were developed under this effort were designed to demonstrate significant future challenges where new technologies and advanced capabilities would be required. Additionally, although the more-detailed scenarios exclusively considered a revolutionary technology future, they covered a broad range of future contingencies. In this way, the scenario portfolio could allow assessments of technology robustness across scenarios to occur as technology assessments are conducted for the full portfolio. Future efforts may consider expanding this scenario portfolio more fully to include variation in the degree of technology development and polarity of actors considered.

Applying the technology road-mapping process for these scenarios necessitated consideration of technologies with high degrees of uncertainty. This deep uncertainty about the future required the use of SMEs to ultimately generate information about technology readiness, impacts, and utility for the scenario studied. The use of SMEs to solicit this information may be required in instances of deep uncertainty, and care should be taken in identifying appropriate expertise. Furthermore, techniques, such as the interview protocol detailed in Appendix B, should be considered to provide a repeatable process for the solicitation of information from SMEs. As more information becomes available about future scenarios and technologies, alternative approaches may be employed such as quantitative modeling and simulation.

Finally, it should be noted that a systematic process should be put in place to assess technologies against future scenarios. The developed technology road-mapping process detailed in this report provides a framework for technology road-mapping activities and offers the structure to repeatably decompose scenarios into sets of challenges, align technologies to challenges, and assess Army modernization priorities against candidate technology solutions.

Recommendations for Future Technology Assessment

To operationalize these findings on technology assessments, the Army should consider implementing technology assessment frameworks to provide structured, repeatable processes for future activities. To support this, the Army should consider the following recommendations:

- The Army should be prepared for a broad range of future contingencies and world states in which technology is diffused and the United States does not have a technological advantage. To support this the Army ought to further develop the portfolio of scenarios considered under technology assessment activities.
- The Army should consider the desired outcomes for future technology assessments to ensure that the appropriate technology concept is applied.
- For future technology road-mapping activities reliant on SME input, the Army should implement a structured process based on best practices to elicit information, such as that provided in Appendix B.
- The Army should leverage technology assessment methods to address modernization priorities and requirement gaps. In conducting further technology assessments, the Army should prioritize the following:
  - Identify techniques to determine the robustness of candidate technology solutions across scenarios.
  - Consider soliciting information from alternative sources such as crowd-forecasting.
  - Continually re-evaluate the current state of knowledge for future technologies, and consider methods integrating quantitative modeling and simulation when sufficient certainty exists to support such activities.

The future is uncertain, and predictions are unlikely to be true. That said, conducting technology assessments in rigorous and robust ways might enable the Army to best prepare itself for future contingencies and operating environments.

Prioritized Scenario Descriptions

We initially developed a list of 20 scenarios—however, engagement with Army stakeholders led us to prioritize five scenarios for more in-depth development. These scenarios were designed according to a template and examples used by AFC and provided as an independent deliverable for Army use in tabletop exercises, technology assessments, and other activities.[1] We also present these five scenarios in a brief narrative form below.

Scenario 1: Battle of the Arctic Depths

Problem statement: How does the Army (as part of the JIIM) operate as part of a combined force in an expeditionary manner, maneuvering long distances in an area characterized by extreme cold weather, difficult terrain, and high-altitude conditions, prevent a peer adversary from solidifying territorial claims?

Global context: Climate change has melted the Arctic region's protective shield of ice and opened its vast lands, seas, and plethora of natural resources to human activity. But states with territorial claims in the Arctic—and key near-Arctic nations—do not agree on how these resources should be divided, with many proffering overlapping claims. China maintains that the region is one of "undetermined sovereignty." Complicating these territorial claims, a severely weakened Russian economy created the conditions for China to purchase swaths of Russia's northern coastline. Russia made this territorial trade in lieu of repaying $10 trillion in infrastructure and

---

[1] These scenario slides were provided to the Army for use in tabletop exercises and therefore are not available to the general public.

technology loans from China extended to Russia, which could not be repaid because of the collapse of global oil prices. China is now using this territory, and its broader policy position that the Arctic's sovereignty is undetermined, to buttress its claims in the Arctic. At play is a range of mineable minerals, including gold, silver, diamond, copper, titanium, graphite, uranium, and other valuable rare earth elements. China's economy is also still heavily reliant on oil, and the region contains between one-fifth and a quarter of the world's untapped fossil-fuel resources (a combination of oil and natural gas).[2]

Road to war: Conflict between the United States and China breaks out over Danish land claims and invocation of NATO's Article 5. Denmark claims an area of 895,000 km[2] extending from Greenland past the North Pole to the limits of the Russian Exclusive Economic Zone. The Danish Claim extends across the North Pole and into Russia's sector.[3] Under the guise of a scientific exploration trip, Chinese forces plant a flag and claim sovereignty over a 300,000 km[2] swath of Arctic territory, all of which is claimed by Denmark. As a result of China's attempts to bolster sovereignty through "facts on the ground," NATO invokes Article 5 to forcibly remove the PLA from the contested area. The United States and its NATO allies thus find themselves in armed conflict with a peer state in the Arctic.

Operational challenges/decisions at play: This scenario surfaced three key challenges for the U.S. Army. First, how to establish an expeditionary presence without a sustained logistics tail in an austere, unpredictable environment, which poses a challenge to supplying fuel and power and equipment maintenance. Second, this scenario highlights the challenge of achieving interoperability with partners. There are eight Arctic nations, five of which are NATO member states (United States, Canada, Denmark, Iceland, and Norway).[4] Third, the environmental conditions in the Arctic present a uniquely hazardous environment for communications. Geostationary satellite links cannot function poleward of 80 degrees because of the curvature

---

2   Maria Smotrytska, "Conquering the North: The Battle for the Arctic," *Modern Diplomacy*, February 3, 2021.

3   United Nations, Division for Ocean Affairs and the Law of the Sea, 2015.

4   The other three Arctic nations are Finland, Russia and Sweden.

of the Earth. Low Earth orbit (LEO) systems (e.g., Iridium satellite constellation) function at the poles, but connectivity can be interrupted, and the current LEO constellations do not provide broadband communication and data transfer is expensive. High frequency (HF) (3 to 30 MHz) radio signals allow long-distance, low-power communications.[5] But, the polar ionosphere can pose a hazardous environment for HF radio signals. Specifically, researchers studying communications in the polar regions note, "the signals often reflect from multiple regions of the ionosphere and by multiple hops with intermediate ground reflections producing multipath effects. As the ionosphere is moving, these signals are also subject to very significant Doppler shifts that add to the complexity of the environment."[6] The combination and unpredictability of these effects make data communications at polar latitudes difficult at times and often only possible at low data rates.

Scenario 2: Gulf War III

Problem statement: How does the Army (as part of the JIIM) fight a counterinsurgency operation against nonstate actors armed with weaponized UAVs that extend their standoff and reach, challenge the ability of U.S. forces to mass, and deny U.S. forces sanctuary?

Global context: The world reached peak oil demand in 2019. Over the subsequent three decades, global demand for oil plummeted as the United States, China, and the European Union—countries which represented roughly half of worldwide oil consumption—transitioned their economies to reliance on renewable sources of energy, particularly hydrogen. Saudi Arabia, previously the world's largest oil exporter, has been particularly affected by these events.[7] Efforts to diversify its revenue base over a three-

---

[5] HF signals propagate via the ionosphere, allowing long distance transmission beyond the horizon.

[6] Michael Prior-Jones, and Mike Warrington, "Digital HF Communications for the Polar Regions—A Low-Cost Alternative to Satellite?" *Geophysical Research Abstracts*, Vol. 12, No. EGU2010-14943, October 2010.

[7] Bassam Fattouh, *Saudi Oil Policy: Continuity and Change in the Era of the Energy Transition,* Oxford Institute for Energy Studies, OIES Paper WPM 81, January 2021, p. 1.

decade period were largely unsuccessful, and the Kingdom's oil revenues still accounted for roughly 85 percent of total government revenues.[8] Government revenues fell by 44 percent from greatly diminished oil sales and were no longer sufficient to underpin the Saudi welfare state of subsidized fuel, water, food, and other essentials.[9] Technology has diffused globally, particularly with respect to autonomous vehicles. Additionally, low-cost, commercially available, portable software defined radios (SDRs) have proliferated, in this case in the form of an app that can be downloaded onto a smartphone, along with open-source code capable of transmitting spoofed GPS signals.[10] The on-demand consumer economy has expanded globally (i.e., Amazon and its imitators provide the world's consumers with ready access to commercial goods).

Road to war: The cost of the Saudi monarchy's cradle to grave welfare state became unsustainable due to the collapse in demand for oil. The bargain between the rulers and its people and military broke. Social unrest in rejection of the monarchy boiled over on December 17, 2050, the 40th anniversary of the Arab Spring. Migrant workers and sectarian groups formed an array of militia groups and were joined by the Saudi armed forces in overthrowing the monarchy in a tumultuous two-week period. The alliance of these groups soon fractured, though, with nationalist, Islamist, and leftist groups all competing for power. A multisided civil war thus broke out across the country. Three of these groups garnered popular support, but the leadership of these groups fled to the United Arab Emirates given the rampant fighting across the country. These individuals remain in the United Arab Emirates as the "Saudi government in waiting," with the United Nations working to broker an end to the civil war and their return to the country. Different groups have threatened to target strategic infrastructure, such as water desalination plants, oil refineries, and power plants if their demands for a position of power in the new Saudi government are not met.

---

[8]   Fattouh, 2021.

[9]   David Fickling, "How Saudi Arabia Can Thrive in a Post-Oil World," *Bloomberg*, April 5, 2021; Katie Paul, "MIDEAST MONEY—Saudi Budget Marks End of Era for Lavish Gulf Welfare Handouts," Reuters, December, 30, 2015.

[10] Center for Advanced Defense Studies, *Above Us, Only Stars: Exposing GPS Spoofing in Russia and Syria*, Washington, D.C., March 26, 2019, p. 13.

The U.S. government resumed its agreement with Saudi Arabia, allowing U.S. forces to permanently station troops at Prince Sultan Air Base southeast of Riyadh. These forces were intended to provide a show of force to deter Iran from further regional adventurism. Now, however, these forces find themselves in the middle of a civil war. A variety of groups—sectarian, labor, military, and mercenaries—are engaged in fighting throughout the country, with the military seemingly having formed an alliance with the Wahhabi sect. Army forces are tasked with ensuring that there is a state for the future government to assume control over once the United Nations–brokered power-sharing agreement is reached.

Operational challenges/decisions at play: This scenario surfaced three key challenges for the Army. First, technologies to improve battlespace awareness, coordination, and force protection have large electronic emissions signatures. These signatures can be exploited by relatively unsophisticated adversaries. For example, the active protection systems on various vehicles to protect against antitank guided missiles and other antitank weapons dramatically increased a unit's signature in the electromagnetic spectrum.[11] The Army needs to be able to manage emissions from electronic systems, especially sensors, to camouflage U.S. positions. Yet, tactics to dampen the emission signature of Army forces limit situational awareness and ability to coordinate.[12] Second, the technical requirements for missile defense to defeat sophisticated aircraft relative to defeating swarms of drones are not symmetrical, and neither are the economics. Adversaries could use armies of small, cheap, plastic drones to monitor friendly movements via electronic emissions and launch electronic and kinetic attacks with kamikaze UAVs, a strategy that would overwhelm existing defenses. Third, diffusion of various technologies could enable nonstate actors, even if unsupported by an external power, to create challenges for Army forces similar to a high-end conflict. For example, SDRs could be used by nonstate actors to mimic authentic GPS satellite signals and spoof U.S. forces with respect to the location of the adversary.

---

[11] Joseph Trevithick, "This Is What Ground Forces Look Like to An Electronic Warfare System and Why It's a Big Deal," *War Zone*, May 11, 2020.

[12] Trevithick, 2020.

Scenario 3: Edison Abroad

Problem statement: How does the Army (as part of a combined international force) operate in an HADR capacity and distribute humanitarian aid across vast distances in a region characterized by limited transportation infrastructure and in close proximity to an adversary state (China)?

- How does the Army accomplish its mission of delivering water and food to the local population, without triggering a military response in China's sphere of influence?
- How does it identify the needs of the region and build local support—at the macro and micro levels—in an environment of misinformation and with limited situational awareness?

Global context: Climate change caused a severe, global freshwater shortage. This environmental reality has been exacerbated in the Southeastern Asian region due to Chinese economic growth and resultant demand for resources to fuel this growth. China views water as a "sovereign" rather than a "shared" resource and has acted accordingly, drawing water from the Mekong River to meet the needs of the Chinese people.[13] Chinese policymakers have taken the view that, "Not one drop of China's water should be shared without China using it first or without making those downstream pay for it."[14] Roughly 200 million people rely on the Mekong River to support farming and fishing, in addition to regular water needs.[15]

Road to HADR mission: Amid a region-wide draught, China holds back large amounts of water upstream on the Mekong River. The Mekong River starts in China and runs through five countries: Cambodia, Laos, Thailand, Burma (Myanmar), and Vietnam. China has a history of holding back large amounts of water, though each time China has dismissed scientific reports of its activities as "groundless." For more than three decades, China has treated

---

[13] Brian Eyler and Courtney Weatherby, *New Evidence: How China Turned off the Tap on the Mekong River*, Washington, D.C.: Stimson Center, April 2020.

[14] Eyler and Weatherby, 2020.

[15] Huilang Tan, "China Could Have Choked Off the Mekong and Aggravated a Drought, Threatening the Lifeline of Millions in Asia," CNBC, April 27, 2020.

data related to water flow and hydropower operations as a state secret. Absent transparency about the extent of its own freshwater holdings, China has been able to propagate a narrative of "shared suffering" due to the drought. This false narrative allowed China to establish common cause for China to deepen its economic cooperation with the countries reliant on the Mekong.[16]

With the water flow reduced for the past six months, amid a region-wide drought, the situation on the ground is dire in all five countries, with heavy crop loss and significantly reduced fish stock. Tens of millions lack access to freshwater. Cambodia, Laos, Thailand, and Vietnam have called on the international community for assistance. China's response to these calls for international intervention have been highly threatening. China insists it has not reduced the flow of water and that reports on the ground are a conspiracy perpetrated by Western governments to damage China's image in the region. China further maintains that if these states require more water, their demands can be accommodated through financial arrangements. China also views the humanitarian efforts as strategic maneuver by the United States, using assistance under the guise to realize the long-held goal of "getting boots on the ground" and establishing a larger, long-term presence in the region. Finally, China has developed the Mekong River into a passage for massive cargo ships to transit from the Yunnan province through these countries to the South China Sea. The passage has been expanded such that it has become an important throughfare for the PLA.[17] These factors mean that this waterway is strategically significant for the Chinese Communist Party (CCP) and PLA. Given the strategic significance of the region, and the hardline position taken by the CCP that foreign governments/militaries should not intervene in its sphere of influence, the United States and its allies are attempting to avoid stumbling into armed conflict with China.

Operational challenges/decisions at play: This scenario surfaced three key challenges for the Army. First, information will be contested, degraded, and denied. The Army will need to be able to identify the needs of the local population and build local support in an environment of misinformation and a lack of "ground truth." Second, avoiding escalation of the humanitar-

---

[16] Eyler and Weatherby, 2020.

[17] Tan, 2020.

ian relief effort to a high-end conflict will require a relatively small footprint that must nonetheless accomplish its mission over a large geographic area. Third, Army forces will face several logistics challenges. U.S. forces will need to be almost entirely self-sufficient given the already dire humanitarian situation. Forces will have to operate without a sustained logistics tail in an austere environment, which poses a challenge to supplying fuel and power and equipment maintenance (albeit limited resource requirements compared with that required in a high-end fight). The availability of roadways and airfields will be unpredictable. Finally, the Army must establish a military presence for an unbounded length of time—however long it takes to deliver the required resources to the local population and implement a sustainable solution for accessing water.

Scenario 4: Fog-of-War Machines

Problem statement: How does the Army (as part of a joint NATO exercise) operate in a deception-dominant environment to reassure allies while signaling resolve to adversary leaders? How can the Army undermine the belief among both partners and the enemy that adversary military deception is effective?

Global context: In 2035, NATO-Russia relations remain icy. U.S. goals relative to Russia are still to deter Russian aggression at all levels and reassure allies and partners of the reliability of U.S. security guarantees. Russia has harnessed AI to optimize deception and undermine adversary situational awareness, resulting in the introduction of automated deception planners, popularly dubbed *fog-of-war machines*. The erosion of NATO awareness of Russian military activity alarms U.S. allies, particularly those whose territory borders Russia's. An imminent U.S.–NATO joint exercise, FOGLIFTER, aims to reassure allies and enhance deterrence by demonstrating that the United States is not intimidated by Moscow's large-scale military deception.

Road to exercise: FOGLIFTER was announced far in advance, allowing the United States and its allies to make extensive preparations and for the Russians to plan their counterexercise. The goals of FOGLIFTER are to conduct counterdeception operations to reduce the effectiveness of Russian fog-of-war machines and improve Blue/Green situational awareness while simultaneously conducting Blue/Green deception operations with the goal of both reducing the effectiveness of Red deception operations and demonstrating NATO deception operations.[18] Additional goals are to reassure NATO allies and other security partners of U.S. commitment and capabilities and to deter and dissuade Russia by reducing Red confidence in the effectiveness of its deception operations. NATO ground forces taking part in FOGLIFTER comprise three armored brigade combat teams (ABCTs) totaling about approximately 40,000 troops; of these, 30,000 are from the United States with remainder from other NATO countries. These ABCTs are distributed across Suwalki Gap and are complemented by simultaneous naval exercise with 45 ships and submarines, 60 aircraft, and 5,500 personnel. All of these forces are assigned to U.S. European Command, which is also the supported command.

Operational challenges/decisions at play: The Russian fog-of-war machines are automated deception planners that work by leveraging the intrinsic hardness of information fusion and appear to be genuinely effective. Although Russian deception operations are planned automatically, many of them are carried out via traditional means of military deception such as inflatable decoys and feints. The effectiveness of Russian deception is greatly enhanced by AI-generated fake traffic signals and new high-tech types of jammers, spoofers, and other counter-ISR systems. The actual disposition of Russian forces is unknown, and U.S.–NATO attempts to reconstruct it find physically impossible, contradictory results (i.e., a Russian unit is reported to be in three places simultaneously). The knowledge quality problems created by fog-of-war machines make it difficult to define operational success, much less measure it. Uncertainty about the disposition of Russian assets increases the difficulty of planning and conducting deception and counterdeception operations. Although Russian forces can exploit fixed facilities in Kaliningrad and the western military district to support their deception operations, NATO forces have to rely much more on temporary and/or mobile facilities.

---

[18] *Blue/Green* refers to host-nation forces (Green) allied with the United States (Blue) when Blue is fighting from another territory.

Scenario 5: Thomas Schelling in the DPRK

Problem statement: How does the Army (in cooperation with other U.S. services and Republic of Korea [ROK] and Japanese forces) use advanced MRADS and SHORADS to reduce political effect of DPRK bombardments and avoid escalation, while also preventing the use of atomic demolition munitions and evacuating at-risk populations?

Global context: The alliance between the United States and the ROK has grown strained because of South Korean perceptions of wavering U.S. commitment. Following the apparent death of Kim Jong-Un, a domestic struggle has apparently broken out between different factions within the DPRK, some of whom seem to believe that their positions will be strengthened if the DPRK makes a significant attack on the south. The Chinese government is ambivalent about these developments: Although Chinese leaders are disinclined to allow the DPRK regime to collapse altogether, they are skeptical of intervening on behalf of the contending North Korean factions. Because of the DPRK's arsenal of weapons of mass destruction (WMDs), it is imperative that escalation be kept under control; however, ROK leaders will feel significant political pressure to mount potentially escalatory retaliation using their own means (such as theater-range ballistic missiles) if damage to the ROK becomes significant. Fortunately, advances in short- and medium-range missile defenses provide the United States with potentially potent tools to help limit such damage and reassure allied leaders. But these systems are not invincible: They have a finite battery depth, and the associated battle management system can only track a few dozen distinct targets in a single sector at a time, being confounded by larger numbers (e.g., decoys and chaff).

Road to war: DPRK forces have started bombarding Seoul and some other more-distant targets in the ROK. U.S. forces already in theater—numbering about 28,500—are being supplemented by additional forces being deployed to total of 33 brigade equivalent for ground forces, while mobilization of ROK forces will bring total ROK forces to 87 brigade equivalent. U.S. Forces Korea and Allied forces are under Combined Forces Command (CFC). Political pressure is building in the ROK demanding potentially escalatory retaliation, but, for the time being, these demands are being kept in check by highly effective new cruise and ballistic missile defenses operated by the United States. However, these defenses are only effective so long as sufficient interceptor missiles/ directed energy assemblies are available when and where they are needed.

The crisis emerged over several months, allowing the ROK and United States to make contingency preparations. Full-scale deployment to the Korean Peninsula only began a few weeks ago, however, so the first additional U.S. brigades are only beginning to arrive at seaports of debarkation in ROK. Although the DPRK assault began with artillery bombardments of Seoul, these quickly abated and were replaced by missile salvos launched from concealed mobile platforms against a wider variety of ROK targets. Intelligence analysis suggests that this restraint is probably a North Korean gambit to increase coercive leverage while reducing incentives for an all-out U.S./ROK counteroffensive against fixed DPRK artillery positions because these might be needed in a later phase of hostilities. The objective of the missile defense operation is the efficient allocation and employment of area and point defense assets to limit the damage caused by DPRK artillery and missile attacks. This factor demands effective emplacement or relocation of missile defense assets, timely and efficient resupply of missile defense units, and prevention of wasteful expenditure of missile defense assets (e.g., interceptors).

Operational challenges/decisions at play: Although U.S. missile defenses appear highly effective at defeating initial North Korean salvos, this may not hold as the adversary refines its tactics and interceptors or equipment are degraded. Only finite missile defense resources are available and misallocating these could enable DPRK saturation tactics. Additionally, missile defense resources need to be conserved to limit damage in an escalation scenario (i.e., DPRK WMD use). Moreover, the missile defense mission competes for resources with general deployment to be prepared for potential escalation contingencies. Blue forces will also prepare to respond to escalation contingencies including adversary WMD employment and possible need to evacuate U.S. citizens and military personnel, an all-out DPRK artillery bombardment of Seoul, and DPRK regime collapse. Although the ROK government has not ordered a general evacuation of Seoul or any other South Korean city, large-scale self-evacuations have taken place, creating logistical obstacles to mobilization and deployment of ROK and U.S. forces in the ROK. Furthermore, although remaining far short of attacking U.S. or partner militaries, China's naval assets are harassing convoys transporting reinforcements to the Korean Peninsula.

Sample Assessment Activity: Technology Matching

This draft protocol outlines an assessment activity aligned to Step 3 through Step 6 of the technology assessment approach detailed in Figure 3.2. The objectives of this assessment activity are to

- gain SME feedback on the overall mission to task framework
- gain SME guidance on a specific action, tack, and set of challenges
- elicit information technology matches, and Army modernization alignment.

The text below in italics is not meant to be read; it is for context only.

Introduction

For the reminder of our time today, we would like to ask you some questions about the scenario, *[enter name of scenario]*. The objective for these questions is (1) gain subject-matter expert (SME) feedback on the overall mission to task framework, (2) gain SME guidance on a specific action, tack, and set of challenges, and (3) elicit information technology matches, and army modernization alignment.

Today we will begin with a set of mission step, objective, subobjective, action, task, challenge. First we will ask you for clarifications on the overall framework, or what might be missing. Then we will move to a technology elicitation exercise. *[If SME has knowledge on this]* We will end with eliciting your thoughts on army modernization alignment.

Clarifications on Overall Framework

- *[If the SME had no questions on the overall framework]* First, I would like to ask your overall thoughts of the framework spanning the mission step, objective, subobjective, action, task, challenge, technology alignment, and Army modernization alignment.
  - What questions do you have about the overall framework or the objectives it is trying to accomplish?
  - What doesn't make sense?
- Next we'd like to focus in on one of the mission steps. For the purposes of today, let us step through the analysis that would occur for an identified set of mission steps, objectives, subobjectives, actions, and tasks. We have identified three options today. *[Show new slide with the three options.]* Which of the following would you like to focus on first?
- Great, we will focus on this one first. *[Show new slide with only one option. Do not show challenges or subsequent columns yet.]* Please read the mission steps, objectives, subobjectives, actions, and tasks. Could you please describe each piece in your own words?
  - *[If not answered]* Could you please describe this **mission step** in your own words?
  - *[If not answered]* Could you please describe this **objective** in your own words?
  - *[If not answered]* Could you please describe this **subobjective** in your own words?
  - *[If not answered]* Could you please describe this **action** in your own words?
  - [If not answered] Could you please describe this **task** in your own words?

Action, Task, and Challenges

- Do you think this is the right task for this action? Why or why not?
- Now, I'd like to walk through the challenges associated with this action and task.
  - What are challenges that you would think would arise?

- *[If not indicated]* What about if [raise point about the scenario]? Would this have the same challenges, or different challenges?
- Our SMEs went through this exercise, and determined the following. *[Click to advance the animations. This should add the challenges text.]* What do you think about these? How do they differ, and how are they the same as what you had identified?
  - *[If same]* Thank you for this input.
  - *[If different]* Thank you for this input. Is there anything else you would like to add to the challenges before, for the purposes of today's discussion, we move to the next exercise?

Technology Matches

- Now, I'd like to walk through the technologies associated with this action and task. What are technologies that might be able to do this action/task and set of challenges?
  - *If subject refers to the Venn Diagram (Figure B.1).* Excellent idea. First, we typically ask SMEs to brainstorm, on the off chance that they would come up with something not on that list. If this is the case, we would add this technology to the list. Since you mention it, let us show the list.
  - *[If subject does not refer to the Venn Diagram]* Thank you. Just as we just did, first, we typically ask SMEs to brainstorm, on the off chance that they would come up with something not on the list previously described in our presentation. Next we would share the list.
- *[Show the Venn Diagram and 19 technologies]* Which of these would be the technologies you would choose for this action/task pairing and set of challenges?
- We also asked other SMEs to identify technologies for this. *[Show results from other SMEs]* What do you think about these? How do they differ, and how are they the same as what you had identified?
  - *[If same]* Thank you for this input.
  - *[If different]* Thank you for this input. Is there anything else you would like to add to the technologies before, for the purposes of today's discussion, we move to the next exercise?

FIGURE B.1
Technology Domains

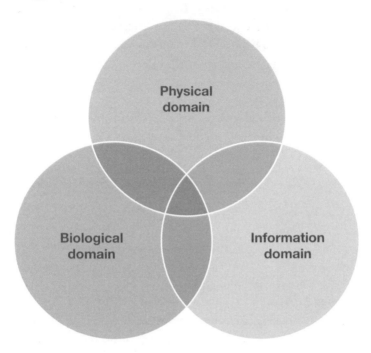

- Next, we would ask a bit more about the technologies. Let us focus on *[enter technology name; hopefully one we are prepared to talk about]*.
  - What is this technology?
  - How would it meet the action/task pairing?
  - How would it or wouldn't it meet the challenges?
- Thank you for telling me a bit more about what you were thinking for this technology.
  - Could you tell me about the status of this technology? *[Prompts included below]* What about:
    - primary investment sector (commercial versus military)
    - leading entity, or entities
    - potential applications, particularly those of interest to DoD
    - time frame of technology development, looking at both evolutionary and revolutionary development
    - potential implementation barriers

- enabling/interdependent technologies
- relevance to specific warfighting functions and cross-functional teams
- near-peer adversaries' level of research and development and investment in the domain.
- Let us stop here for the moment. You have raised a number of insightful data that we will work to combine with other SMEs to flesh out this technology.

Army Modernization Alignment

- *[If the SME has knowledge on this]* What can you tell us about Army modernization alignment?

Close Out and Next Steps

Thank you for your time today. As mentioned previously, we will use your responses to help us refine the overall process. Ideally, these refinements would allow us to improve our research discussions with SMEs, and those data would be used to inform the overall object of the research. Now, I'd like to turn it back to the moderators for closing and next steps.

Battle of the Arctic Depths Technology Research

The following appendix provides a more-detailed assessment of each of the candidate technologies identified for the Battle of the Arctic Depths scenario discussed in Chapter Three.

Lubricants

Lubricants reduce friction, wear, and energy consumption and are subsequently required to operate every piece of machinery. There are four types of lubricants—oil, grease, penetrating lubricants, and dry lubricants—and each is optimized for a different use case. Oil and grease are the two most widely used, while penetrating and dry lubricants have more-specific use cases, the former being used to break up rust or debris and the latter being a useful alternative to avoid attracting dust and dirt with wet lubricants.[1]

In cold weather, viscosity becomes of great concern, specifically as it relates to lubricants. Most greases and base oils can withstand temperatures dipping as low as –10°C. However, when temperatures get much colder, lubricants become significantly less viscous and start to stiffen, which can cause equipment to seize up and malfunction. As a result, in recent years, R&D related to lubricants, particularly in extremely cold conditions (e.g., the Arctic), has increased, resulting in a host of lubricants and greases with

---

[1] S. P. Srivastava, "Chapter 1: Introduction: Lubricant Basics," in S. P. Srivastava, ed., *Developments in Lubricant Technology*, Hoboken, N.J.: John Wiley & Sons, Inc., August 2014.

ideal properties. Both the automotive and maritime industries are investing in this space—particularly the latter, given the presence of Arctic shipping routes.

To select the best lubricant for a given environment, it is important to understand a lubricant's viscosity index and pour point. The viscosity index is a unitless measure of the change in viscosity relative to the change in temperature—the higher the viscosity index, the less likely viscosity is affected by changes in temperature, which makes a given lubricant ideal for cold weather scenarios. The pour point is the temperature at which oil begins to stiffen and will no longer flow. Table C.1 shows a list of common base oils along with their viscosity indexes and pour points. A lubricating grease is typically a combination of at least one base oil, a thickener, and an additive. Table C.2 shows the pour points of various combinations of base oils (rows) and thickeners (columns), with thickeners consisting of lithium soap (Li), aluminum soap (Al), sodium soap (Na), calcium soap (Ca), barium soap (Ba), and PTFE (or silicone).[2]

In addition to increasing interest from several public sectors, the Army has invested in the development of new lubricants through GVSC's Fuels and Lubricants Research Facility (located at Southwest Research Institute in

TABLE C.1
Cold Weather Properties of Base Oils

Base Oil	Viscosity Index	Pour Point
Refined mineral oil	80–100	Poor low temperature behavior
Synthesized PAOs (polyalphaolefin)	150–250	−40 to −50°C
Esters (natural or synthesized)	140–175	−50°C or below
PAGs (polyalkylene glycol)	150–270	−40 to −50°C
PFPE (perfluorinated polyether)	50–350	Some can go as low as −70°C
Silicone oils	—	−50°C or below

SOURCE: Klüber Lubrication, undated.

---

[2] Klüber Lubrication, "Lubricant Challenges in Extreme Cold Environments," webpage, undated.

TABLE C.2.

Cold Weather Properties of Greases

Oil Type	Li	Al	Na	Ca	Ba	PTFE
Mineral oil	−10° to −35°C	−10 to −30°C	−20° to −30°C	−10° to −30°C	−10° to −20°C	N/A
PAO	−25° to −50°C	−25 to −50°C	N/A	−35° to −50°C	−35° to −50°C	−20° to −50°C
Esters	−40° to −70°C	N/A	N/A	−20° to −40°C	N/A	N/A
PAG	−30° to −50°C	N/A	N/A	N/A	N/A	−20° to −30°C
PFPE	N/A	N/A	−25° to −40°C	N/A	N/A	−20° to −70°C
Silicone	−40° to −50°C	N/A	−30° to −60°C	N/A	N/A	−40° to −70°C

SOURCE: Klüber Lubrication, undated.
NOTE: Al = aluminum soap; Ba = barium soap; Ca = calcium soap; Li = lithium soap; Na = sodium soap; N/A = not applicable; PTFE = Polytetrafluoroethylene (or silicone).

Texas) and the Armament Research, Development, and Engineering Center at Picatinny Arsenal in New Jersey. Testing of an SCPL—a synthetic, all-season, fuel-efficient, heavy-duty engine oil developed at the GVSC research facility—was focused specifically on the Arctic. Additionally, FEGO, which is an axle differential gear oil developed at the TARDEC, was designed to be used on very low-friction surfaces, such as ice. As of the writing of this report in 2021, testing of these lubricants in the Arctic and in other cold-weather environments has been successful.

Although choosing the right lubricant is a critical consideration with far-reaching impact, particularly in cold-weather climates, given the level of investment coming from multiple sectors along with ongoing research efforts, we believe that this is a problem readily navigated in the technology space.

Quantum Technologies

Quantum science combines elements of mathematics, computer science, engineering, and physical science to study photons and electrons—the smallest particles of matter and energy. Unlike classical mechanics, where objects exist in a specific place and time, the existence of an object is probabilistic in quantum mechanics. At the subatomic level, objects can also act as both a wave and a particle, depending on whether the object is being measured—because the objects being measured are so small, the simple act of measuring the object fundamentally alters its nature.[3]

Quantum technologies can be broken down into three applications—computing, communications, and sensing. The most-heralded applications include cryptography and cryptanalysis, enabling precision timing and navigation in GPS-denied environments, identification of moving masses underwater and underground structures, the end of stealth through improved sensors, and pinpointing electric field sensors and communications receivers. Quantum computing will pose a distinct threat to systems that rely on public key cryptography, rendering all present and future data vulnerable. Quantum communications would provide the ability to conduct communications that cannot be broken, improving the security of communications against interception and eavesdropping.[4] Magnetometers and gravitometers could be used to detect underground vacancies (e.g., bunkers) or dense materials (e.g., nuclear materials). Although magnetometers are used to detect locations of shipwrecks and other objects under the sea or under the Earth, quantum magnetic sensors are being explored, with some experts claiming these sensors could offer a million-fold increase in sensitivity.[5] Quantum sensors could also contribute to improved inertial navigation systems, allowing submarines to operate without active sonar (thereby

---

[3]   Jonathan P. Dowling and Gerald J. Milburn, "Quantum Technology: The Second Quantum Revolution," *Philosophical Transactions of the Royal Society A*, Vol. 361, No. 1809, 2003.

[4]   Tucker, 2017.

[5]   Stephen Battersby, "Core Concept: Quantum Sensors Probe Uncharted Territories, From Earth's Crust to the Human Brain," *Proceedings of the National Academy of Sciences of the United States of America*, Vol. 116, No. 34, 2019.

complicating the adversary's ability to find and attack submarines).[6] Two quantum imaging techniques—known as ghost imaging and quantum radar—improve intelligence via enhanced image resolution and signal sensitivity. Specifically, the former would allow for imaging despite atmospheric impediments (e.g., smoke, clouds),[7] while the latter is capable of detecting low-visibility targets against very noisy backgrounds.[8]

Not unlike many others, the Army is investing heavily in quantum technologies, both by funding research efforts and establishing new agencies, positions, and programs dedicated to this space. Some examples include the work spearheaded by DEVCOM and the Army Research Laboratory's Center for Distributed Quantum Information, focused on improving sensing and methods to build quantum communication networks,[9] and the research hub recently stood up by the Army Research Office and National Security Agency's Laboratory for Physical Sciences, which is part of the National Quantum Initiative—a coordinated federal plan to accelerate and advance quantum technology-centered studies and applications.[10]

It is worth noting that, although garnering significant interest from virtually all sectors and possessing the ability to be truly transformative in the battlespace, even experts wildly disagree on the feasibility of these technologies within certain time frames. In short, quantum advances may perpetually be on the horizon. In the next 20 years, few (if any) of these technologies are likely to advance to the point where they are capable of being deployed in support of combat operations. As an example, quantum technologies often require cryogenic temperatures and the absence of vibrations in ways that would be difficult (and, if even possible, very costly) to replicate outside a

---

[6] Cameron, 2019; Kramer, 2014.

[7] Ronald E. Meyers, Keith S. Deacon, and Yanhua Shih, "Turbulence-Free Ghost Imaging," *Applied Physics Letters*, Vol. 98, No. 11, 2011; Peter A. Morris, Reuben S. Aspden, Jessica E. C. Bell, Robert W. Boyd, and Miles J. Padgett, "Imaging with a Small Number of Photons," *Nature Communications*, Vol. 6, No. 5913, 2015.

[8] Seth Lloyd, "Enhanced Sensitivity of Photodetection via Quantum Illumination," *Science*, Vol. 321, No. 5895, 2008, p. 1463.

[9] South, 2021; U.S. Army Combat Capabilities Development Command, Army Research Laboratory Public Affairs, 2021; Zhong et al., 2021.

[10] Vincent, 2021.

highly controlled laboratory environment. Quantum communications and computing remain highly experimental, whereas quantum sensors may not be sufficiently small, light, low power, or cost-effective to operate in a battlefield environment in 2040.

Biotechnology

Although current definitions vary, the term *biotechnology* remains a broad term for the exploitation of biological processes, organisms, cells, and cellular components to develop new technologies or products.[11] It is a growing, ever-changing field garnering interest from all sectors.

There are many different potential biotechnology applications that are relevant in this Arctic scenario. In short, biotechnology has the ability to enhance warfighting materiel and systems, optimizing warfighter health and performance and military medicine. External devices, such as exoskeletons, augment an individual's strength, making it easier to carry and assemble billeting at high elevations and in deep snow.[12] Research in pharmaceuticals, such as modafinil and other drugs, could provide both cognitive and physical enhancements ranging from improved attention, concentration, learning, and memory to enhanced energy production and improved physical performance.[13] Research in materials has the potential to revolutionize the wound healing process along with the design and function of future systems.[14] Other advances could contribute to the ability to remove contami-

---

[11] James Carafano and Andrew Gudgel, *National Security and Biotechnology: Small Science with a Big Potential*, Washington, D.C.: Heritage Foundation, 2007.

[12] Joseph Camberato, "New Exoskeleton Gives Construction Workers Super-Human Strength," National Business Capital, last updated January 16, 2019.

[13] Modafinil is a medication that promotes wakefulness. Maddalena Mereu, Antonello Bonci, Amy H. Newman, and Gianluigi Tanda, "The Neurobiology of Modafinil as an Enhancer of Cognitive Performance and a Potential Treatment for Substance Use Disorders," *Psychopharmacology*, Vol. 229, No. 3, 2013.

[14] B. Li, J. M. Davidson, and S. A. Guelcher, "The Effect of the Local Delivery of Platelet-Derived Growth Factor from Reactive Two-Component Polyurethane Scaffolds on the Healing in Rat Skin Excisional Wounds," *Biomaterials*, Vol. 30, No. 20, 2009; Ramasatyaveni Geesala, Nimai Bar, Neha R. Dhoke, Pratyay Basak, and Ami-

nants and salts from water to make it safe for consumption and general use, the miniaturization and general improvement of medical devices and diagnostics, and even the development of biological energy sources.[15] Truthfully, the potential use cases for biotechnology in any scenario are countless given the exceedingly broad variety of technologies and applications that fall under this heading.

The Army is also investing heavily in this space, as are the sister services. Synthetic biology in particular remains a priority research area within the Army and across the broader DoD. The Army Research Lab has the Transformational Synthetic Biology for Military Environments Essential Research Program, focused on self-assembling and self-healing materials, active camouflage, and on-demand production of small-quantity materials.[16] The ERDC has a history of genetically modifying microorganisms for biological detection and contaminant (e.g., oil spill, munition) degradation.

Many of these biotechnologies are presently at a maturation level that allow for deployment in the field in their current state, though further R&D will only continue to improve these technologies while also creating new ones. We recommend that the Army continue investing in biotech-

---

tava Das, "Porous Polymer Scaffold for On-Site Delivery of Stem Cells—Protects from Oxidative Stress and Potentiates Wound Tissue Repair," *Biomaterials*, Vol. 77, 2016; Marc Bohner, "Resorbable Biomaterials as Bone Graft Substitutes," *Materials Today*, Vol. 13, No. 1–2, 2010; Pamela Habibovic and Klaas de Groot, "Osteoinductive Biomaterials–Properties and Relevance in Bone Repair," *Journal of Tissue Engineering and Regenerative Medicine*, Vol. 1, No. 1, January–February 2007; K. Yoshimura, H. Eto, H. Kato, K. Doi, and N. Aoi, "In Vivo Manipulation of Stem Cells for Adipose Tissue Repair/Reconstruction," *Regenerative Medicine*, Vol. 6, No. 6S, 2011.

[15] Burak Derkus, "Applying the Miniaturization Technologies for Biosensor Design," *Biosensors and Bioelectronics*, Vol. 79, No. 15, 2016; Iain McConnell, Gonghu Li, and Gary W. Brudvig, "Energy Conversion in Natural and Artificial Photosynthesis," *Chemistry & Biology*, Vol. 17, No. 5, 2010; J. Barber, "Photosynthetic Energy Conversion: Natural and Artificial," *Chemical Society Reviews*, Vol. 38, No. 1, January 2009; Ku Kalyanasundaram and Michael Gräetzel, "Artificial Photosynthesis: Biomimetic Approaches to Solar Energy Conversion and Storage," *Current Opinion in Biotechnology*, Vol. 21, No. 3, June 2010.

[16] U.S. Army Combat Capabilities Development Command, Army Research Laboratory Public Affairs, "Army Scientists Explore Synthetic Biology Potential," June 24, 2019.

nology while simultaneously looking to industry and academia for further advances in the space.

Space Transport for Earth P2P

The space sector is developing P2P space technology for transportation from a point to another point on Earth. By utilizing suborbital space flight, space P2P launch vehicles would allow transporting humans and cargo to anywhere on Earth at a speed much faster than the current air transportation.

The concept of using space to travel to terrestrial destinations dates to the 1950s with the Army's Project Adam, and the concept has been revived from time to time since then.[17] The previous concepts of space P2P never materialized, but the latest effort from the commercial space sector has been more promising with developments and tests of such launch systems ongoing. What make the latest effort different from the previous ones are the continued reduction in launch cost, successes in developing and operating reusable launch systems, and private sector–led development as opposed to military-led.[18]

The Arctic scenario involves transporting 11,600 soldiers and relevant equipment and sustainment supplies to an austere place without established infrastructure as one of the main challenges. Space P2P technology presents several advantages, along with the potential payload capacity of 100 tons, for solutions to the challenges in the Arctic scenario in comparison with traditional air travel.[19]

Of many advantages, first and foremost is the significant reduction in travel time. Although the technology is in development and flight demonstrations have not occurred, one prominent company promises major

---

[17] Examples include Project ROMBUS (Reusable Orbital Module-Booster and Utility Shuttle) in the 1960s and Project Hot Eagle in the early 2000s.

[18] James D. Powers, Jan Osburg, Jeffrey Brown, Thao Liz Nguyen, Moon Kim, Kristin F. Lynch, and Thomas Hamilton, *Assessing Global Rapid Mobility Using Next-Generation Space Vehicles*, Santa Monica, Calif.: RAND Corporation, Not available to public.

[19] Michael Sheetz, "The Pentagon Wants to Use Private Rockets Like SpaceX's Starship to Deliver Cargo Around the World," CNBC, June 4, 2021.

international routes in around 30 minutes.[20] Another advantage is that that the use of suborbital flight trajectories eliminates the need to acquire diplomatic overflight access because sovereignty in space is not recognized in the international community. Some space launch vehicles also do not require a runway and have the potential to land and take-off in austere zones.

Nonetheless, achieving operational capability for space P2P depends on significant developments in related technologies. The commercial sector is working on improving technologies to enhance propulsion and landing technologies; to reduce costs of launch, recovery, and refurbishment; and, perhaps most importantly, to meet safety levels for commercial viability. Although rapid prototyping and testing are ongoing, significant uncertainty exists in the availability period for operational capability (from the late 2020s to late 2030s). Also, the accessibility of the technology will heavily depend on the success of commercial P2P business. Without a commercial market, DoD would have to pay a substantial price to use space P2P as a mode of transportation.

From the military, the U.S. Space Force and the AFRL are leading the Rocket Cargo Vanguard program to leverage the commercial market's development of space P2P capabilities. Although the Army currently has no plans or capabilities for this technology and will rely on traditional logistics/resupply capabilities over land, air, and water for the foreseeable future, the Rocket Cargo Vanguard program can support the U.S. Transportation Command's logistics capabilities.

We recommend the Army conduct use case studies with focus on landing and take-off in austere and extreme environments (e.g., the Arctic), ensure that access to space P2P system is available to the Army, and prepare for integration so early adoption of the technology is possible when it becomes ready for operational capability.

Air Vehicles–Future VTOL

In austere environments, such as the Arctic, that lack infrastructure and runways for conventional airplanes, air vehicles with VTOL capability can be

---

[20] SpaceX, "Earth to Earth Time Comparisons," webpage, undated.

an advantageous transportation mechanism. VTOL is a not a new technology. Legacy helicopters and rotorcrafts, such as Chinook and Black Hawks, have served as VTOL air vehicles for the military for several decades. Many current UAVs also have VTOL capability.

Key related technologies, such as rotors, propulsion, situational awareness with AI and advanced sensors, avionics, communication, and advanced composite materials, are currently making advances. Significant improvements in these technologies should take place over the next decade to optimize size, weight, and power of VTOL aircrafts while increasing payload capacity, range, and speed. Both the defense and commercial industries are advancing these technologies for next-generation VTOL vehicles.

A notable trend in the technology is the commercial sector's development of electric-VTOL (eVTOL) aircrafts with electric propulsion systems. Numerous companies across the world are developing eVTOL for various uses including regional air mobility, urban air mobility, and urban cargo delivery. The capabilities of the eVTOLs in development vary, with travel distance from 20 to 150 miles and cruise speed from 60 to 200 miles per hour.[21] Although the sector aims to deliver first generation eVTOL passenger vehicles by mid to late 2020s, many challenges related to regulations, infrastructure, and safety lie ahead for mainstream commercial adoption.[22]

Army Modernization efforts categorize VTOL in five capability sets, from light to heavy, through the FVL program.[23] Among the capabilities, the Army's current focus is on light and medium transport VTOL platforms. The medium capability will be able to transport 12 soldiers and have external payload of 8,000 to 10,000 pounds.[24] The FVL program anticipates operational capability of the earliest system to likely be in the early 2030s.[25]

---

[21] Woodrow Bellamy, "10 eVTOL Development Programs to Watch in 2021," *Avionics*, February/March 2021.

[22] Bellamy, 2021.; Robin Lineberger, Aijaz Hussain, and Vincent Rutgers, "Change Is in the Air. The Elevated Future of Mobility: What's Next on the Horizon?" *Deloitte Insights*, June 3, 2019.

[23] Gertler, 2021.

[24] Steven A. Yeadon, "The Impact of the Future Long-Range Assault Aircraft on United States Army Air Assaults—Part 2," *Aviation Digest*, Vol. 8, No. 3, 2020.

[25] Gertler, 2021.

Capability sets 4 (heavy lift) and 5 (ultra-heavy lift) are more relevant to the transportation challenges presented in the Arctic scenario considering the significant number of troops, equipment, and supplies involved. Capability set 4 can carry around 30 passengers with 20,000 pounds of external payload, and capability set 5 can carry up to around 50 passengers with more than 30,000 pounds of external payload.[26] The time frame for the initial operational capability for the two capability sets are uncertain, with unofficial estimates going out to the 2060s. One company in the commercial sector, however, is developing an eVTOL vehicle that has the capacity to transport 40 passengers or 10,000 pounds of cargo, with the plan to offer commercial cargo transportation service by the mid-2020s.[27]

Although VTOL technology is certainly attractive for a mission similar to the one presented in the Arctic scenario, development is lacking on the heavy and ultra-heavy payload vehicles that would be relevant for the transportation challenges presented in the Arctic scenario. Considering that the readiness time frame for the light and medium capability sets are in the 2030s, it is unlikely for the heavier capability sets to be available in the 2030s. As the Army modernization effort is focused on the first three VTOL capability sets as of the writing of this report in 2021, we recommend that the Army continues to engage with the commercial sector for the potential availability of heavy and ultra-heavy payload VTOLs.

AWS

With the Arctic scenario staged in the 2030s, the Army can expect AWS to be available for a variety of applications. DoD defines *autonomous weapon system* as "a weapon system that, once activated, can select and engage tar-

---

[26] Jeffrey A. Drezner, Parisa Roshan, and Thomas Whitmore, *Enhancing Management of the Joint Future Vertical Lift Initiative*, Santa Monica, Calif.: RAND Corporation, RR-2010-OSD/JS, 2017.

[27] Luke Dormehl, "Forget Drone Taxis. This Startup is Building a 40-Seat Drone Bus," *Digital Trends,* June 7, 2021.

gets without further intervention by a human operator."[28] DoD also considers those systems that allow human supervision for overriding operations. Current policies separate offensive AWS, also known as LAWS, from defensive AWS. Focusing on the task of coordinating offensive missions to expel Red (Mission Step 5), this section describes the current state of development of several key AWS for offensive operations, and their future potentials.

The international community does not have consensus on the development of LAWS, mainly because of the ethical considerations around such weapon systems. For example, the current U.S. policy "does not prohibit the development or employment of LAWS,"[29] while China called for a ban on the use of, but not the development nor production of, LAWS.[30] Many countries also have supported the ban on the use and development of LAWS.

Despite the lack of international consensus, developments in the required technology are continuing. Many technologies for LAWS derive from technologies in nonlethal autonomous systems, which are not subject to the same ethical criticism at home and abroad. The following is a nonexhaustive list of such systems, with operational examples, that share technologies required for LAWS:

- UASs: Autonomous aerial mobility
  - examples: General Atomics Aeronautical Systems Reaper MQ-9A, Kratos XQ-58A Valkyrie, and AFRL's Skyborg Program
- loitering munition: autonomous aerial loitering and search and autonomous swarming capabilities
  - examples: Raytheon Coyote and Israeli Harpy
- unmanned ground vehicles: autonomous terrestrial mobility
  - example: Estonian THeMIS

---

[28] Department of Defense Directive 3000.09, *Autonomy in Weapon Systems*, Washington, D.C.: U.S. Department of Defense, November 21, 2012; incorporating change 1, May 8, 2017.

[29] Kelley M. Sayler, "Defense Primer: U.S. Policy on Lethal Autonomous Weapons," Congressional Research Service, IF11150, last updated December 1, 2020.

[30] Mary Wareham, "Stopping Killer Robots: Country Positions on Banning Fully Autonomous Weapons and Retaining Human Control," Human Rights Watch, August 10, 2020.

- autonomous defense systems: autonomous detection, identification, and strike of incoming aerial threat
  - examples: Patriot Missile System, Aegis Combat System, and Israeli Iron Dome
- robotic sentry weapons: autonomous detection and identification of terrestrial threats
  - examples: South Korean SGR-A1 and Israeli Sentry Tech.

Many of the systems are capable of fully autonomous operations that do not require human interaction. The aerial vehicles above can fly, loiter, and identify targets autonomously. Furthermore, as seen in the use of Libya's Government of National Accord in 2020, today's drones can be programmed to strike targets without human intervention.[31] Terrestrial vehicles are not as advanced as aerial vehicles, but robotic combat vehicles (RCVs) that can conduct autonomous terrestrial mobility in various environmental conditions are in development.

Future development of AWS will be based on evolutionary improvements of the current technologies. For example, advances in AI, machine learning, and sensors will improve object recognition. Enhancements in advanced materials, actuators, and motors will improve maneuverability and sustainability. Developments in battery and computer chips will better the efficiency of AWS.

Because of the contention over the ethical considerations of such systems, the United States has policies that require some sort of human-in-the-loop operation for offensives systems, and the capability to override autonomous decisions for defensive systems. For example, Aegis is considered a human-supervised AWS that can identify and strike targets, but human operators can override the operation.[32] DoD policies also will require UASs and RCVs with strike and fire capabilities to include human decisionmak-

---

[31] United Nations Security Council, *Final Report of the Panel of Experts on Libya Established Pursuant to Security Council Resolution 1973 (2011)*, New York, March 8, 2021.

[32] Robert O. Work, "Principles for the Combat Employment of Weapon Systems with Autonomous Functionalities," Center for New American Security, April 28, 2021.

ing for engagement.[33] Therefore, the availability and the accessibility of AWS for the Army would depend less on the technological factors and more on the policy decisions.

The Army is actively pursuing technologies related to AWS. Investment in technology development includes research in autonomous and semiautonomous vehicles and weapon systems, although human judgment for targeting and firing capabilities is still the main operating concept. For example, the Army has published a strategy for the present and near future of UAS.[34] The Army also has numerous UASs already in operation and acquired loitering munitions, with further development and acquisition in process. Next Generation Combat Vehicle CFT, FVL CFT, Assured PNT CFT, Network CFT are several efforts within the Army that are focused on enhancing the technologies related to AWS.

---

[33] AFRL, "Skyborg," webpage, undated; Kris Osborn, "The Army Could Soon Use Robots as Armed Weapons (And Much More)," *National Interest*, December 2, 2019.

[34] U.S. Army, Unmanned Aircraft Systems Center of Excellence, *Eyes of the Army: U.S. Army Unmanned Aircraft Systems Roadmap 2010–2035*, Fort Rucker, Ala., April 2010.

Abbreviations

AI	artificial intelligence
AFC	Army Futures Command
AFRL	U.S. Air Force Research Laboratory
AWS	autonomous weapon systems
CCP	Chinese Communist Party
CFT	Cross-Functional Team
COP	common operational picture
DEVCOM	U.S. Army Combat Capability Development
DoD	U.S. Department of Defense
DPRK	Democratic People's Republic of Korea
ERDC	U.S. Army Engineer Research and Development Center
eVTOL	electric-vertical take-off and landing
FEGO	fuel-efficient gear oil
FVL	future vertical lift
GVSC	Ground Vehicle Systems Center
HADR	humanitarian assistance and disaster relief
HF	high frequency
ISR	intelligence, surveillance, and reconnaissance
JIIM	Joint Interagency, Intergovernmental, and Multinational
LAWS	lethal autonomous weapon systems
NATO	North Atlantic Treaty Organization
P2P	point-to-point
PLA	People's Liberation Army
PNT	positioning, navigation, and timing

R&D	research and development
RCV	robotic combat vehicle
ROK	Republic of Korea
SCPL	single common powertrain lubricant
SDR	software defined radio
SME	subject-matter expert
TAA	tactical assembly area
TARDEC	U.S. Army Tank Automotive Research, Development, and Engineering Center
UAS	unmanned aerial system
UAV	unmanned aerial vehicle
VTOL	vertical take-off and landing
WFF	warfighting functions
WMD	weapon of mass destruction

References

AFC—*See* Army Futures Command.

AFRL—*See* Air Force Research Laboratory.

Air Force Research Laboratory, "Skyborg," webpage, undated. As of August 3, 2021:
https://afresearchlab.com/technology/vanguards/successstories/skyborg

Army Futures Command, *Future Operational Environment: Forging the Future in an Uncertain World, 2035–2050*, Austin, Tex., AFC PAM 525-2, 2019.

Banta, David, "What Is Technology Assessment?" *International Journal of Technology Assessment in Health Care*, Vol. 25, No. S1, July 2009, pp. 7–9.

Barber, J., "Photosynthetic Energy Conversion: Natural and Artificial," *Chemical Society Reviews*, Vol. 38, No. 1, January 2009, pp. 185–196.

Battersby, Stephen, "Core Concept: Quantum Sensors Probe Uncharted Territories, From Earth's Crust to the Human Brain," *Proceedings of the National Academy of Sciences of the United States of America*, Vol. 116, No. 34, 2019, pp. 16663–16665.

Bellamy, Woodrow, "10 eVTOL Development Programs to Watch in 2021," *Avionics*, February/March 2021. As of July 28, 2021:
http://interactive.aviationtoday.com/avionicsmagazine/
february-march-2021/10-evtol-development-programs-to-watch-in-2021/

Blais, Curtis L., *Unmanned Systems Interoperability Standards*, Monterey, Calif.: Naval Postgraduate School, NPS-MV-16-001, September 2016.

Bohner, Marc, "Resorbable Biomaterials as Bone Graft Substitutes," *Materials Today*, Vol. 13, No. 1–2, 2010, pp. 24–30.

Bütschi, Danielle, Rainer Carius, Michael Decker, Søren Gram, Armin Grunwald, Petr Machleidt, Stef Steyaert, and Rinie van Est, "The Practice of TA: Science, Interaction, and Communication," in Michael Decker, Miltos Ladikas, Susanne Stephan, and Friederike Wütscher, eds., *Bridges Between Science, Society and Policy*, Berlin: Springer, 2004, pp. 13–55.

Camberato, Joseph, "New Exoskeleton Gives Construction Workers Super-Human Strength," National Business Capital, last updated January 16, 2019. As of August 26, 2021:
https://www.nationalbusinesscapital.com/
new-exoskeleton-gives-construction-workers-superhuman-strength/

Cameron, Alan, "Quantum Magnetometer Senses Its Place," *GPS World*, May 8, 2019. As of August 23, 2021:
https://www.gpsworld.com/quantum-magnetometer-senses-its-place/

Carafano, James, and Andrew Gudgel, *National Security and Biotechnology: Small Science with a Big Potential*, Washington, D.C.: Heritage Foundation, 2007.

Castells, Manuel, *The Information Age: Economy, Society, and Culture,* Vol. 1, *The Rise of the Network Society*, Hoboken, N.J.: Wiley-Blackwell, 1996.

Center for Advanced Defense Studies, *Above Us, Only Stars: Exposing GPS Spoofing in Russia and Syria*, Washington, D.C., March 26, 2019. As of August 24, 2021:
https://www.c4reports.org/aboveusonlystars

Chamberlin, Paul Thomas, *The Cold War's Killing Fields: Rethinking the Long Peace*, New York: Harper Collins, 2019.

Copeland, Thomas E., eds., *The Information Revolution and National Security*, Carlisle Barracks, Pa.: U.S. Army War College, August 2000.

Department of Defense Directive 3000.09, *Autonomy in Weapon Systems*, Washington, D.C.: U.S. Department of Defense, November 21, 2012; incorporating change 1, May 8, 2017.

Derkus, Burak, "Applying the Miniaturization Technologies for Biosensor Design," *Biosensors Bioelectronics*, Vol. 79, May 15, 2016, pp. 901–913.

Dormehl, Luke, "Forget Drone Taxis. This Startup is Building a 40-Seat Drone Bus," *Digital Trends*, June 7, 2021. As of July 29, 2021:
https://www.digitaltrends.com/features/flying-drone-bus-vtol/

Dowling, Jonathan P., and Gerald J. Milburn, "Quantum Technology: The Second Quantum Revolution," *Philosophical Transactions of the Royal Society A*, Vol. 361, No. 1809, June 2003.

Drezner, Jeffrey A., Parisa Roshan, and Thomas Whitmore, *Enhancing Management of the Joint Future Vertical Lift Initiative*, Santa Monica, Calif.: RAND Corporation, RR-2010-OSD/JS, 2017. As of March 31, 2022:
https://www.rand.org/pubs/research_reports/RR2010.html

Encyclopedia.com, "Constructive Technology Assessment," webpage, undated. As of March 30, 2022:
https://www.encyclopedia.com/science/
encyclopedias-almanacs-transcripts-and-maps/
constructive-technology-assessment

Eyler, Brian, and Courtney Weatherby, *New Evidence: How China Turned off the Tap on the Mekong River*, Washington, D.C.: Stimson Center, April 2020.

Fattouh, Bassam, *Saudi Oil Policy: Continuity and Change in the Era of the Energy Transition*, Oxford Institute for Energy Studies, OIES Paper WPM 81, January 2021.

Fickling, David, "How Saudi Arabia can Thrive in a Post-Oil World," *Bloomberg*, April 5, 2021. As of August 24, 2022:
https://www.bloomberg.com/opinion/articles/2021-04-05/how-saudi-arabia-can-thrive-in-a-post-oil-world

Gaddis, John Lewis, *The Long Peace: Inquiries into the History of the Cold War*, New York: Oxford University Press, 1987.

Garcia, Marie L., and Olin H. Bray, *Fundamentals of Technology Roadmapping*, Albuquerque, N.M.: Strategic Business Development, Sandia National Laboratories, SAND97-0665, 1997.

Geesala, Ramasatyaveni, Nimai Bar, Neha R. Dhoke, Pratyay Basak, and Amitava Das, "Porous Polymer Scaffold for On-Site Delivery of Stem Cells— Protects from Oxidative Stress and Potentiates Wound Tissue Repair," *Biomaterials*, Vol. 77, January 2016, pp. 1–13.

Genus, Audley, "Rethinking Constructive Technology Assessment as Democratic, Reflective, Discourse," *Technological Forecasting and Social Change*, Vol. 73, No. 1, 2006, pp. 13–26.

Gertler, Jeremiah, "Army Future Vertical Lift (FVL) Program," Congressional Research Service, IF11367, last updated July 13, 2021. As of March 31, 2022:
https://sgp.fas.org/crs/weapons/IF11367.pdf

Gonzales, Daniel, Timothy R. Gulden, Aaron Strong, and William Hoyle, *Cost–Benefit Analysis of Proposed California Oil and Gas Refinery Regulations*, Santa Monica, Calif.: RAND Corporation, RR-1421-DIR, 2016. As of March 29, 2022:
https://www.rand.org/pubs/research_reports/RR1421.html

Gordon, Theodore J., and Olaf Helmer-Hirschberg, *Report on a Long-Range Forecasting Study*, Santa Monica, Calif.: RAND Corporation, P-2982, 1964. As of August 18, 2021:
https://www.rand.org/pubs/papers/P2982.html

Grunwald, Armin, "Technology Assessment: Concepts and Methods," in Anthonie Meijers, ed., *The Handbook of the Philosophy of Science, Philosophy of Technology and Engineering Sciences*, Amsterdam, Netherlands: North-Holland, 2009, pp. 1103–1146.

Habibovic, Pamela, and Klaas de Groot, "Osteoinductive Biomaterials— Properties and Relevance in Bone Repair," *Journal of Tissue Engineering and Regenerative Medicine*, Vol. 1, No. 1, January–February 2007, pp. 25–32.

Hsia, Pei, Jayarajan Samuel, Jerry Gao, David Kung, Yasufumi Toyoshima, and Cris Chen, "Formal Approach to Scenario Analysis," *IEEE Software*, Vol. 11, No. 2, 1994, pp. 33–41.

Huss, William R., "A Move Toward Scenario Analysis," *International Journal of Forecasting*, Vol. 4, No. 3, 1988, pp. 377–388.

Kahen, Goel, "Strategic Development, Technology Transfer and Strategic Technology Assessment in Changing Environments," *Proceeding of the First International Conference on Dynamics of Strategy*, Guildford, United Kingdom, April 11–12, 1996.

Kahn, Herman, and Anthony J. Weiner, *The Year 2000: A Framework for Speculation on the Next Thirty-Three Years*, New York: Macmillan Publishers, 1967.

Kalyanasundaram, Ku, and Michael Gräetzel, "Artificial Photosynthesis: Biomimetic Approaches to Solar Energy Conversion and Storage," *Current Opinion in Biotechnology*, Vol. 21, No. 3, June 2010, pp. 298–310.

Kang, Daekook, Wooseok Jang, Hyeonjeong Lee, and Hyun Joung No, "A Review on Technology Forecasting Methods and Their Application Area," *International Journal of Mechanical, Industrial and Aerospace Sciences*, Vol. 7, No. 4, 2013, pp. 591–595.

Klüber Lubrication, "Lubricant Challenges in Extreme Cold Environments," webpage, undated. As of August 26, 2021:
https://www.klueber.com/us/en/company/newsroom/news/
lubricant-challenges-in-extreme-cold-environments/

Konrad, Kornelia, Arie Rip, and Verena Schulze Greiving, "Constructive Technology Assessment—STS for and with Technology Actors," *European Association for the Study of Science and Technology*, Vol. 36, No. 3, 2017.

Kosow, Hannah, and Robert Gaßner, *Methods of Future and Scenario Analysis: Overview, Assessment, and Selection Criteria*, Bonn, Germany: German Development Institute and Deutsches Institut für Entwicklungspolitik, 2008.

Kramer, David, "DARPA Looks Beyond GPS for Positioning, Navigating, and Timing," *Physics Today*, Vol. 67, No. 19, 2014.

Le Guin, Ursula K., *Hainish Novels and Stories*, Vol. 1, New York: Library of America, 2017.

Lempert, Robert J., "Robust Decision Making (RDM)," in Vincent Marchau, Warren Walker, Pieter Bloemen, and Steven Popper, eds., *Decision Making Under Deep Uncertainty*, Cham, Switzerland: Springer, 2019, pp. 23–51.

Lempert, Robert J., Steven W. Popper, and Steven C. Bankes, *Shaping the Next One Hundred Years: New Methods for Quantitative, Long-Term Policy Analysis*, Santa Monica, Calif.: RAND Corporation, MR-1626-RPC, 2003. As of March 30, 2022:
https://www.rand.org/pubs/monograph_reports/MR1626.html

Li, B., J. M. Davidson, and S. A. Guelcher, "The Effect of the Local Delivery of Platelet-Derived Growth Factor from Reactive Two-Component Polyurethane Scaffolds on the Healing in Rat Skin Excisional Wounds," *Biomaterials*, Vol. 30, No. 20, July 2009, pp. 3486–3494.

Lineberger, Robin, Aijaz Hussain, and Vincent Rutgers, "Change Is in the Air. The Elevated Future of Mobility: What's Next on the Horizon?" *Deloitte Insights*, June 3, 2019. As of July 28, 2021: https://www2.deloitte.com/us/en/insights/focus/future-of-mobility/ evtol-elevated-future-of-mobility-summary.html

Lloyd, Seth, "Enhanced Sensitivity of Photodetection via Quantum Illumination," *Science*, Vol. 321, No. 5895, 2008, pp. 1463–1465.

Mattis, James, *Summary of the 2018 National Defense Strategy of the United States of America: Sharpening the American Military's Competitive Edge*, Washington, D.C.: U.S. Department of Defense, 2018.

Mazarr, Michael J., Ashley L. Rhoades, Nathan Beauchamp-Mustafaga, Alexis A. Blanc, Derek Eaton, Katie Feistel, Edward Geist, Timothy R. Heath, Christian Johnson, Krista Langeland, Jasmin Léveillé, Dara Massicot, Samantha McBirney, Stephanie Pezard, Clint Reach, Padmaja Vedula, and Emily Yoder, *Disrupting Deterrence: Examining the Effects of Technologies on Strategic Deterrence in the 21st Century*, Santa Monica, Calif.: RAND Corporation, RR-A595-1, 2022. As of May 10, 2022: https://www.rand.org/pubs/research_reports/RRA595-1.html

McConnell, Iain, Gonghu Li, and Gary W. Brudvig, "Energy Conversion in Natural and Artificial Photosynthesis," *Chemical & Biology*, Vol. 17, No. 5, May 28, 2010, pp. 434–447.

Mereu, Maddalena, Antonello Bonci, Amy H. Newman, and Gianluigi Tanda, "The Neurobiology of Modafinil as an Enhancer of Cognitive Performance and a Potential Treatment for Substance Use Disorders," *Psychopharmacology*, Vol. 229, No. 3, 2013, pp. 415–434.

Meyers, Ronald E., Keith S. Deacon, and Yanhua Shih, "Turbulence-Free Ghost Imaging," *Applied Physics Letters*, Vol. 98, No. 11, 2011.

Morris, Peter A., Reuben S. Aspden, Jessica E. C. Bell, Robert W. Boyd, and Miles J. Padgett, "Imaging with a Small Number of Photons," *Nature Communications*, Vol. 6, No. 5913, 2015.

Osborn, Kris, "The Army Could Soon Use Robots as Armed Weapons (And Much More)," *National Interest*, December 2, 2019. As of August 4, 2021: https://nationalinterest.org/blog/buzz/ army-could-soon-use-robots-armored-weapons-and-much-more-101157

Paul, Katie, "MIDEAST MONEY— Saudi Budget Marks End of Era for Lavish Gulf Welfare Handouts," *Reuters*, December 30, 2015.

Postma, Theo J. B. M., and Franz Liebl, "How to Improve Scenario Analysis as a Strategic Management Tool?" *Technological Forecasting and Social Change*, Vol. 72, No. 2, 2005, pp. 161–173.

Powers, James D., Jan Osburg, Jeffrey Brown, Thao Liz Nguyen, Moon Kim, Kristin F. Lynch, and Thomas Hamilton, *Assessing Global Rapid Mobility Using Next-Generation Space Vehicles*, Santa Monica, Calif.: RAND Corporation, Not available to the general public.

Prior-Jones, Michael, and Mike Warrington, "Digital HF Communications for the Polar Regions—A Low-Cost Alternative to Satellite?" *Geophysical Research Abstracts*, Vol. 12, No. EGU2010-14943, October 2010.

Quist, Jaco, and Philip Vergragt, "Past and Future of Backcasting: The Shift to Stakeholder Participation and a Proposal for a Methodological Framework," *Futures*, Vol. 38, No. 9, 2006, pp. 1027–1045.

Richards, C. E., R. C. Lupton, and J. M. Allwood, "Re-framing the Threat of Global Warming: An Empirical Causal Loop Diagram of Climate Change, Food Insecurity and Societal Collapse," *Climatic Change*, Vol. 164, No. 49, 2021.

Rostker, Bernard D., Charles Nemfakos, Henry A. Leonard, Elliot Axelband, Abby Doll, Kimberly N. Hale, Brian McInnis, Richard Mesic, Daniel Tremblay, Roland J. Yardley, and Stephanie Young, *Building Toward an Unmanned Aircraft System Training Strategy*, Santa Monica, Calif.: RAND Corporation, RR-440-OSD, 2014. As of March 31, 2022: https://www.rand.org/pubs/research_reports/RR440.html

Ryan, Thomas, and Vikram Mittal, "Potential for Army Integration of Autonomous Systems by Warfighting Function," *Military Review*, September–October 2019.

Sayler, Kelley M., "Defense Primer: U.S. Policy on Lethal Autonomous Weapons," Congressional Research Service, IF11150, last updated December 1, 2020.

Schot, Johan, and Arie Rip, "The Past and Future of Constructive Technology Assessment," *Technological Forecasting and Social Change*, Vol. 54, No. 2–3, 1997, pp. 251–268.

Schwartz, Peter, *The Art of the Long View: Planning for the Future in an Uncertain World*, New York City: Penguin Random House, 1996.

Sciubba, Jennifer Dabbs, "Demography and Instability in the Developing World," *Orbis*, Vol. 56, No. 2, 2012, pp. 267–277.

Sheetz, Michael, "The Pentagon Wants to Use Private Rockets Like SpaceX's Starship to Deliver Cargo Around the World," CNBC.com, June 4, 2021. As of July 27, 2021: https://www.cnbc.com/2021/06/04/us-military-rocket-cargo-program-for-spacexs-starship-and-others.html

Smits, R., and J. Leyten, *Technology Assessment: Waakhond of Speurhond (Technology Assessment: Watchdog or Tracker? Towards an Integral Technology Policy)*, Vrije Universiteit Amsterdam, Kerckebosch, Zeist, Vol. 28, 1991.

Smotrytska, Maria, "Conquering the North: The Battle for the Arctic," *Modern Diplomacy*, February 3, 2021. As of August 24, 2021:
https://moderndiplomacy.eu/2021/02/03/
conquering-the-north-the-battle-for-the-arctic/

Sofuoğlu, Emrah, and Ahmet Ay, "The Relationship Between Climate Change and Political Instability: The Case of MENA Countries (1985:01–2016:12)," *Environmental Science and Pollution Research*, Vol. 27, April 2020, pp. 14033–14043.

South, Todd, "Quantum Breakthroughs Help Army, Air Force Advance Supercomputing," *Army Times*, April 6, 2021.

SpaceX, "Earth to Earth Time Comparisons," webpage, undated. As of July 26, 2021:
https://www.spacex.com/human-spaceflight/earth/index.html

Srivastava, S. P., "Chapter 1: Introduction: Lubricant Basics," in S. P. Srivastava, ed., *Developments in Lubricant Technology*, Hoboken, N.J.: John Wiley & Sons, Inc., August 2014, pp. 3–6.

Sutton, Walter S., *Cyber Operations and the Warfighting Functions*, Carlisle Barracks, Pa.: U.S. Army War College, March 2013.

Tan, Huilang, "China Could Have Choked Off the Mekong and Aggravated a Drought, Threatening the Lifeline of Millions in Asia," CNBC.com, April 27, 2020. As of March 31, 2022:
https://www.cnbc.com/2020/04/28/china-choked-off-the-mekong-which-worsened-southeast-asia-drought-study.html

Thomas, Charles W., "Strategic Technology Assessment, Future Products and Competitive Advantage," *International Journal of Technology Management*, Vol. 11, No. 5-6, January 1996, pp. 651–666.

Tran, Thien A., "Review of Methods and Tools Applied in Technology Assessment Literature," *PICMET '07— 2007 Portland International Conference on Management of Engineering & Technology*, Portland, Ore., August 5–9, 2007.

Tran, Thien A., and Tugrul Daim, "A Taxonomic Review of Methods and Tools Applied in Technology Assessment," *Technological Forecasting and Social Change*, Vol. 75, No. 9, 2008, pp. 1396–1405.

Trevithick, Joseph, "This Is What Ground Forces Look Like to An Electronic Warfare System and Why It's a Big Deal," *The Drive*, May 11, 2020. As of March 31, 2022:
https://www.thedrive.com/the-war-zone/33401/this-is-what-ground-forces-look-like-to-anelectronic-warfare-system-and-why-its-a-big-deal

Tripp, Robert S., Kristin F. Lynch, John G. Drew, and Robert G. DeFeo, *Improving Air Force Command and Control Through Enhanced Agile Combat Support Planning, Execution, Monitoring, and Control Processes*, Santa Monica, Calif.: RAND Corporation, MG-1070-AF, 2012. As of March 31, 2022:
https://www.rand.org/pubs/monographs/MG1070.html

Tsai, Huei-Ting, and Tzu-Ming Liu, "Effects of Global Climate Change on Disease Epidemics and Social Instability Around the World," Asker, Norway, Human Security and Climate Change International Workshop, June 21–23, 2005. As of March 31, 2022:
http://www.harep.org/Agriculture/HTingTsai.pdf

Tucker, Patrick, "China has a Breakthrough in Spy-Proof Quantum Communications," *Defense One*, November 9, 2017. As of August 23, 2021:
https://www.defenseone.com/technology/2017/11/china-has-breakthrough-spy-proof-quantum-communications/142415/

United Nations, "2019 Revision of World Population Prospects," webpage, undated. As of August 25, 2021:
https://population.un.org/wpp/

United Nations, Division for Ocean Affairs and the Law of the Sea, "Commission on the Limits of the Continental Shelf (CLCS), Outer Limits of the Continental Shelf Beyond 200 Nautical Miles from the Baselines: Submissions to the Commission: Submission by the Kingdom of Denmark," November 2, 2015. As of March 18, 2021:
https://www.un.org/depts/los/clcs_new/submissions_files/submission_dnk_76_2014.htm

United Nations Security Council, *Final Report of the Panel of Experts on Libya Established Pursuant to Security Council Resolution 1973 (2011)*, New York, March 8, 2021.

U.S. Army Combat Capabilities Development Command, Army Research Laboratory Public Affairs, "Army Scientists Explore Synthetic Biology Potential," June 24, 2019. As of August 24, 2021:
https://www.army.mil/article/223495/army_scientists_explore_synthetic_biology_potential

U.S. Army Combat Capabilities Development Command, Army Research Laboratory Public Affairs, "Army-Funded Research Lays Groundwork for Future Quantum Networks," March 11, 2021. As of March 31, 2022:
https://www.army.mil/article/244139

U.S. Army, Unmanned Aircraft Systems Center of Excellence, *Eyes of the Army: U.S. Army Unmanned Aircraft Systems Roadmap 2010–2035*, Fort Rucker, Ala., April 2010. As of March 31, 2022:
https://apps.dtic.mil/sti/pdfs/ADA518437.pdf

Van Den Ende, Jan, Karel Mulder, Marjolijn Knot, Ellen Moors, and Philip Vergragt, "Traditional and Modern Technology Assessment: Toward a Toolkit," *Technological Forecasting and Social Change*, Vol. 58, No. 1–2, May–June 1998, pp. 5–21.

Van Eijndhoven, Josée C. M., "Technology Assessment: Product or Process?" *Technological Forecasting and Social Change*, Vol. 54, No. 2–3, February–March 1997, pp. 269–286.

Vincent, Brandi, "NSA, Army Launch 'Qubit Collaboratory' to Advance Quantum Information Science," NextGov.com, May 4, 2021. As of August 24, 2021:
https://www.nextgov.com/emerging-tech/2021/05/nsa-army-launch-qubit-collaboratory-advance-quantum-information-science/173808/

Wareham, Mary, "Stopping Killer Robots: Country Positions on Banning Fully Autonomous Weapons and Retaining Human Control," Human Rights Watch, August 10, 2020. As of March 31, 2022:
https://www.hrw.org/report/2020/08/10/stopping-killer-robots/country-positions-banning-fully-autonomous-weapons-and#_ftn265

Work, Robert O., "Principles for the Combat Employment of Weapon Systems with Autonomous Functionalities," Center for New American Security, April 28, 2021. As of August 3, 2021:
https://www.cnas.org/publications/reports/proposed-dod-principles-for-the-combat-employment-of-weapon-systems-with-autonomous-functionalities

Yeadon, Steven A., "The Impact of the Future Long-Range Assault Aircraft on United States Army Air Assaults—Part 2," *Aviation Digest*, Vol. 8, No. 3, 2020, pp. 16–26.

Yoshimura, K., H. Eto, H. Kato, K. Doi, and N. Aoi, "In Vivo Manipulation of Stem Cells for Adipose Tissue Repair/Reconstruction," *Regenerative Medicine*, Vol. 6, No. 6S, November 2011, pp. 33–41.

Zhong, Youpeng, Hung-Shen Chang, Audrey Bienfait, Etienne Dumur, Ming-Han Chou, Christopher R. Connor, Joel Grebel, Rhys G. Povey, Haoxiong Yan, David I. Schuster, and Andrew N. Cleland, "Deterministic Multi-Qubit Entanglement in a Quantum Network," *Nature*, Vol. 590, 2021, pp. 571–575.